ENVIRONMENTAL BIOTECHNOLOGY

ENVIRONMENTAL BIOTECHNOLOGY

By

Dr. P.R. Yadav
Lecturer
Department of Zoology
D.A.V. College
Muzaffarnagar (U.P.)

&

Dr. Rajiv Tyagi
Department of Zoology
M.M. College
Modi Nagar (U.P.)

D P H

DISCOVERY PUBLISHING HOUSE
NEW DELHI-110002

Reprinted - 2019

First Published - 2006

ISBN: 978-81-8356-071-9

Environmental Biotechnology

Published by:

DISCOVERY PUBLISHING HOUSE PVT. LTD.

4383/4B, Ansari Road, Darya Ganj

New Delhi-110 002 (India)

Phone: +91-11-23279245, 43596064-65

Fax: +91-11-23253475

E-mail: discoverypublishinghouse@gmail.com

sales@discoverypublishinggroup.com

web: www.discoverypublishinggroup.com

Printed at:

Infinity Imaging Systems

Delhi

Preface

Biotechnology is a multidisciplinary field which has been considered in several National Development Programmes as one of the strategic areas, involving edge technologies and promoting new bio-industries as source of a considerable amount of new products with high impact in agriculture, the food and chemical industry and also in the health sector. Among the various subfields with biotechnology, the one dealing with environmental issues is currently one of the major demand and development. Environmental biotechnologies are increasingly being viewed as a major weapon against environmental damage. The present title "Environmental Biotechnology" deals with such information at various levels, from introductory to advanced.

This volume is aimed at providing information at various levels, from introductory to advanced aspects, including general concepts of environmental policy, applications and specific case studies related to both environmental biotechnology and cleaner bioprocesses. It is also aimed at benefiting readers of different backgrounds, including engineers, consultants, scientists and decision makers working for industry, government, academia and social or private organizations.

In the preparation of this book large number of books and research papers have been consulted. So no authenticity is claimed.

The author wishes to express his deepest appreciation to the many people who have contributed in one way or the other in the preparation of this title.

The author express is gratitude to Mr. Wasan and staff of M/s Discovery Publishing House for their whole hearted co-operation in the publication of this book.

The author tried hard to be accurate and upto date in statement and realises the impossibility of completely avoiding errors therefore, the author will greatly will greatly appreciate having his attention called to any questionable statement.

Authors

Contents

1

INTRODUCTION

The quality of life on earth is to the equal quality of the environment sustaining it and is disturbed by pollution. The effect of "misplaced resources" generates unfavourable conditions, which threaten the fabric of biosphere. Even after associating pollution with the advancement of human civilization, it has no such concern as it was in the last few decades. As we have stepped into the millenium without the essential ecological wisdom, this may surely result in endangering the quality of life of our future generations.

The seizure on the environment has come through the increased pollution pressure, which lead to over exploitation of land, air and water resources and also provoking the ecological balance and implanting various *xenobiotic* (till unknown to nature) substances into the environment, which affects its self-purification and carrying capacity.

These problems have drawn global attention in recent years to find out the ways to sustain and manage environment. It has challenged the environmentalists, scientists, technologists, politicians and also socio-economic reformers. There arises a question of *sustainable development* has appeared which require judicious use of nature without compromising on its capacity to meet the future needs. This results for new outlooks towards developmental activities and technologies, so that eco-health is maintained alongside conservation of invaluable resources. Man has taken just a couple of centuries to destruct them whereas the nature needed thousands of years to create resources, which supports life.

There has been a greater scientific attempts however has changed its direction and technical approach to ensure great human welfare. The emerging science of biotechnology is very essential tool as it provides new approach for understanding, managing, preserving and also to restore environmental quality. Biotechnological methodologies can be used to know the well being of ecosystems, transformation of pollutants into harmless substances, generate biodegradable materials from renewable resources and to bring up ecofriendly manufacturing and disposal processes. All these aspects come under the study of '*Environmental Biotechnology*".

The science of biotechnology is known since long and the concepts of alcoholic beverages, waste water treatment are some examples. With the development of modern biology, based on the know-hows in molecular genetics, cell biology, ecology, mechanisms of microbial metabolism and bio-chemistry, put together with process technology, it gives high promise for human welfare.

Utilizing the handy scientific tool, biotechnology has been defined differently by people. In simple it is a technology which has biomaterial and biological principles to yield beneficiary products to human requirement. Already it is benefited in health care, agriculture and also food industries, which on the environmental side it has added a new ways in pollution biosurveillance, bio-sensing and bio-abatement with the use of genetically modified organisms, with advanced metabolic efficiency to "scavenge" pollutants degrade lignocellulosic components of waste and other complex recalcitrant chemicals. These helps in the minimization of pollution load in the environment. Electronic principles are also used to pollution sensing technology employing immobilized living products such as enzymes, proteins and DNA as sensing agents in biosensors.

Biotechnology has given better understanding of mechanisms of pollutant action at cellular and molecular levels so that fair manipulations could be done during technological process. The concept of *biomethylation* of toxic metals, their effects and *biomagnification* are also benefited. The former indicate the formation of organo-metallic complexes in the environment due to microbial actions and such a conjugate in certain cases (e.g. CH_3–Hg) could be more toxic than the inorganic form of mercury. In the biomagnification process it has been noticed that the concentration of some toxicants gets magnified several folds in organisms through the food chain than their concentrations in environment, e.g., DDT concentration in dolphins is several times more than in the sea water or that of Cd^+ is 50 times more in mollusks than in the surrounding medium.

Pollutant → Phytoplankton and plants → Fishes → Higher animals

With notable achievements of this technology include the clean up of water and land areas polluted with petroleum products through 'superbugs', vast potential in this arena is still to be tapped, namely bioreclamation and rehabilitation of polluted aquatic reservoirs and land areas fouled by pollutants, including toxic metals from industrial effluents.

'Bioindicators' help in pollution monitoring through lab experiments and *microcosm* tests, which would support extrapolation of data in case of *ecological equivalents*. In many cases, damaging effects of pollutants in minute concentrations, which may be below the range of conventional physiochemical techniques, can be reasonably resolved through the biotech methodologies (*toxicity testing*). These known 'hidden' injuries may provide an early signal for the danger which can be avoided through timely remedial measures.

Apart from the benefits of the system, environmental biotechnology can provide cost effective methodologies with reasonable reliability and may prove to be a boon for small-scale industries in developing countries. Its integration with non-biological techniques may provide to be more fruitful in the long run.

Environmental Biotechnology in the 21st Century

The problem of worse of environmental health is discussed both by environmentalists and politicians and calls for early solution. Biotechnology play a significant role in protecting and rehabilitating the environment. The problem now is not restricted to national geographic boundaries but has turned into a global one. Close collaboration between scientists and engineers is essential for development of bioremedial process technology. For eg., gasoline contaminated water is now known to be bio-remediated. But aeration required to stimulate aerobic microbial activity, strips the hydrocarbon from liquid and it exchanges a water pollution problem for an air pollution. For recalcitrant chemicals, use of co-metabolism and genetically engineered microbial strains may often be required within treatment processes of a bio-reactor. This requires considerable ingenuity to solve difficulty and a close co-operation, which would indicate how and in which direction biotechnology should develop to address some of urgent problems over next decades.

Modern physico-chemical technologies have high price tags and may not always achieve decontamination upto a desired level. While environmental biotechnology can not only do same job in a harmless and cost effective way, it can also visualize the pollution impact at a much lower level as its parameters can go even upto molecular levels of living target systems. The only thing that is needed is more of research and development in scale up of processes. Although biotechnology has already shown promises in environmental arena through possibility of development of 'natural' pesticides, biofertilizers, genetically altered disease resistant plants, biodegradable plastics, and improved bioscavenging of pollutants through GEMs, it has not received due attention it deserves. It is hoped that through proper R & D and sincere effort and co-operation of all concerned, this new technology in early part in this millennium would make this planet more habitable for present and future generations, as the quality of life is linked to overall quality of our environment.

The Biotechnological Research Sub-committee (BRS) of National Science & Technical Council of USA and the biotechnology department of Government of India have put emphasis on certain 'Thrust Areas' of research and development in environmental biotechnology for coming decade for restoration of environmental health. These could be broadly grouped as :

1. Detection, monitoring and remediation of toxic substances in soil, water and air,
2. Biocontrol of disease, pests and weeds,
3. Successful biodegradation and biotransformation of xenobiotics, including genetic upgrading of biodegraders,
4. Recycling of wastes and conversion into energy,
5. Restoration ecology for proper balance of structure and function of ecosystem
6. Development of database and informatics and establishment of EPAs as district levels, if possible.

The priority of research focus on following aspects :

Microbes and other Bioscavengers

Their genetics, ecology and physiology data bases to be developed. The possibilities of genetic manipulation for upgrading catabolic activities and sorption potential should be looked into. The fate and activities of 'improved' degraders in the environmental need constant monitoring.

Monitoring of Environmental Quality

1. Water quality surveillance and aerosol biomonitoring. For this rapid, reproducible and inexpensive techniques and kits to be developed.

Otherwise the very purpose of environmental monitoring will be defeated.

2. Development of highly sensitive biosensors and membrane bioprobes for on-line monitoring of dangerous substances, toxic metals and pathogens.
3. Development of non-specific molecular receptors to identify wastes from various industries.

Restoration of Environmental Quality

Application of biotechnology in restoration of contaminated soil, sub-surface and surface waters offers several advantages over conventional techniques. It includes:

1. Screening of degradative microbe for specific xenobiotics like phenol, halogenated aromatics, pesticides, heavy metals and their organo compounds.
2. *In situ* restoration with genetically manipulated scavengers.
3. Engineering of plants for stress tolerance, hyperaccumulation and detoxification of metals.
4. Biotransformation of fungal aflatoxins.
5. Development of bio-pesticides substituting chemical ones.
6. Marine sediments complex are poorly known, here dynamics of microbial enzyme activity has to be worked.
7. Removal of radionuclides by biosorption, eg., Selenium and Uranium by *Geobactor*.
8. Biodiversity conservation which is likely to restore ecological balance.

Resource Recovery

1. Bioleaching of minerals from wastes needs to be augmented through more basic research.
2. Microbial bio-surfactant and bio-polymers need to be developed.
3. Resource conservation and waste minimization.
4. Strain development of methanotrophs for degasification of coal seams and wet lands.
5. Generation of energy and industrial raw materials from wastes, including delignification of agricultural and forest wastes.

Laboratory to Field Evalulation

1. Development of suitable *microcosm* and *mesocosm* tests before full scale field trials.

2. To evaluate long-term efficacy of natural bioremediation and restoration in specific alloted areas.

Bio-safety and Risk Containment

1. Appropriate measures and guidelines for release and regular monitoring of GEMs and their interactions with indigenous microfloras is absolutely necessary. Here UNEP guidelines may be followed.
2. Gene transfer inhibition from one strain of microbe to other has to be worked out. Knowledge of anti-sense RNA could be a boon in this, as it blocks the genetic information flow from DNA-RNA to protein.

It can be said, according to think tanks in the scientific world, that 21st century will be an era of both information technology and biotechnology giving new direction to Human Civilization.

2

MICROBES AND ENVIRONMENT

This chapter is for non-biologist who need to have unique idea about types and groups of microbial organisms often encountered during various biotechnological programs related to environmental issues. Microbes that largely seen in nature (air, water and soil) play different part in matter cycles. Even though they pollute environment, they help to improve its quality. The knowledge of their work in natural conditions and genetically change state is now suitably employed to improve our degrading environment, including waste care and other forms of pollution like refactored xenobiotic chemicals of anthropogenic origin.

In general, microbes take an important position in the development of modern biology as an applied science. It is now unknown to study various biological phenomenon without the knowledge of microbial groups. Their cells being simple serve as attractive models of biological structure and function, metabolism and energetics, genetics and evolutionary processes. The easy and fast methods of their multiplication help bringing up successive generations within limited space and time for any biotechnological operation. Out of the different groups of microbes, bacteria and viruses have been considered to be very important tools in *genetic engineering*.

MICROBIAL GROUPS

As the name says, these organisms can be studied only with the help of microscopes. They have forms like bacteria, fungi, algae and protozoa of which the first 3 are of plant kingdom and the fourth belongs to animal group. Apart from this, there are special classes, which are more simpler and include viruses, mycoplasmas and rickettsias. They are considered to be *ultramicrobes* and identified only with help of electron microscopes. They cannot be grouped under plant or animal world and have put aside as *microtatobiotes*. These ultramicrobes may be a cellular/non-cellular as *viruses*, or cellular with very simple cell structure with no cell wall as in *mycoplasmas* and *rickettsias*. They are infective in nature.

Microbes like bacteria and others having different primitive single celled structures have been grouped under *protists*. Again, on the basis of the nature of nucleus, they have been classified as *procaryotes* (nucleus with non-membrane cover) and *eucaryotes* (nucleus with different surrounding membrane). Bacteria are considered as *procaryotes* while fungi, algae and protozoa come and *eucaryotes*.

Based on their metabolism, microbes may be *parasites* (lives within host cells of plants and animals including man). *Saproplytes* those grow on dead organic matter or *free-living* in nature and carry on different roles in the biogeochemical cycles.

Though microorganisms cause different types of diseases in other living organisms including human (Table 2.1) yet most of them are non-pathogenic. It is this group which is used in number of technological processes including environmental management.

Among the various microbial groups, bacteria is the huge and we are more concerned with regard to their nature, activities and useful aspects in biotechnology. In the subsequent sections this would be studied in detail than other groups.

VIRUSES

Basically, viruses means 'poison'. They were first recognized in 1892 by Iwanovsky as 'infectious filtrable agents', as their physical and chemical nature was then unknown. They were really an enigma till the discovery of electron microbes. They could pass through filters, suggesting their much minute structure than bacteria.

Viruses are total parasites on living hosts and show their living features as reproduction units with host cells. They cannot be refined in media and grown in lab. Outside the host cells, they are insert chemical crystals and are considered to be unusual simpler microbes and represent a borderline between the living and non-living.

The body is non-cellular/acellular, i.e., any type of cell structure is made of nucleo-protein molecule where the protein form the 'coat'and nuclei acid establish the 'core'. This core may be DNA as in animal and bacterial viruses, or RNA as in plant viruses. Outer part of the coat, may be spikes like projections as in Herpes and Influenza virus.

Table 2.1. Some common microbial pathogens of man

Group	*Pathogen*	*Disease*
I. **Viruses**	Influenza virus	Influenza
	Polio virus	Poliomyelitis
	Pox virus	Small pox
	Hepatitis virus	Jaundice
	Rabies virus	Hydrophobia
	HIV	AIDS
	Onco virus	Cancer
	Herpes virus	Skin disease
	Paramyxo virus	Measles and Mumps
II. **Bacteria**	*Salmonella*	Typhoid
	Vibrio	Cholera
	Mycobacterium	T.B.
	Corynebacterium	Diptheria
	Shigella	Blood dysentery
	Pnewmoocci	Pneumonia
	Yersinia	Plague
	Clostridium	Tetanus
	Neisseria	Gonorrhoea
	Treponema	Syphilis
	Staphylococcus	Wounds
III. **Fungi**	*Microporum*	Epidemic Ringover of scalp.
	Trichophyton	Infection of hair, scalp, feet and nails.
	Trichoporon	Infection of hair and scalp.
	Candida	Infection of vaginal mucous membrane, and mouth disease (Thrus).
	Aspergillus	Tokelan (ring worm like) disease, lung infection.
	Sporothrix	Subcutaneous nodules, parti-cularly lymph nodes.

The disease caused by members of different fungal groups are mostly restricted to skin infection and are called dermatomycoses.

IV. **Protozoa**	*Entamoeba*	Amoebiasis
	Balantidium	Balantidiasis (intestinal ulcer)
	Giardia	Giadiasis (Diaorrhea)
	Trypanosoma	Sleeping sickness
	Trichomonas	Vaginitis
	Leishmania	Kala azar
	Plasmodium	Malaria

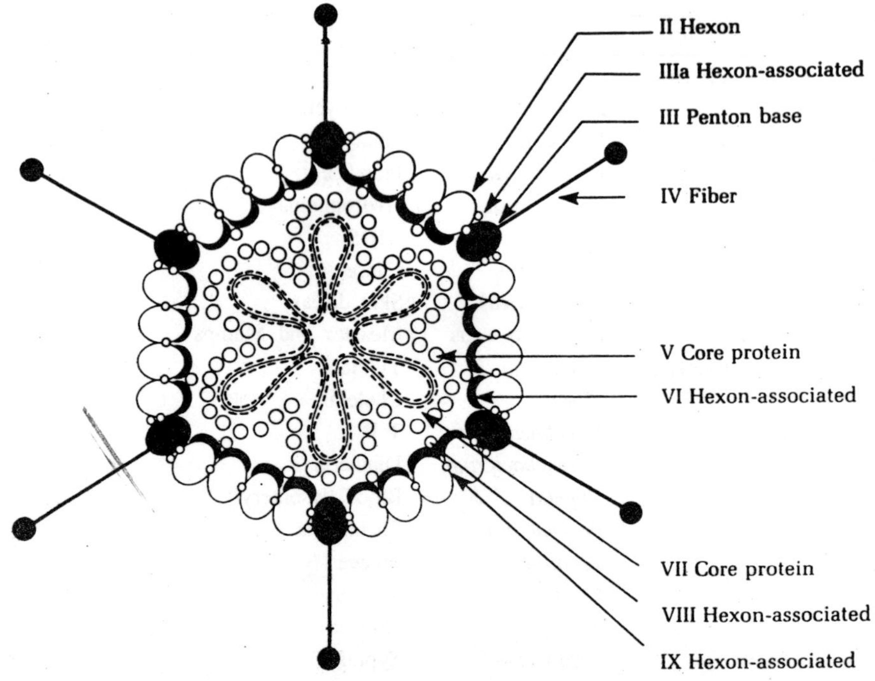

Fig. 2.1. Model of adenovirus.

The size ranges from 20 mm as in Polio virus to 300 mm as in Rabies and Small Pox virus. Shape-wise they may have oval, cuboidal, rods, polygon or tadpole like shapes (as in bacterial virus).

The bacteria are infected by *Bacteriophages*, whereas blue-green algae and fungi are infected by *Cyanophages* and *Mycophages* respectively. Bacteria infective virus are of 2 types of namely, *virulent* when it kills the bacterium (called *lysis*) and moderate when its nucleic acid gets combined with bacterial chromosome and multiplies with it. The presence of virus in a host tissue can be seen when the tissue extract is spread on agar medium and different zones of viruses are seen. This is named as *Plagues*.

Viruses in Drinking Water

All warm blooded animals get into enteric viruses, find their way to aquatic medium. Human enteri viruses found in the environment for more time and many of them can survive water treatment. They can be found both in surface and ground waters. To find out the virus in water, it is done either by plague formation or by cytopathologic effects in tissue culture. Some may be identified by immunofluorescent technique.

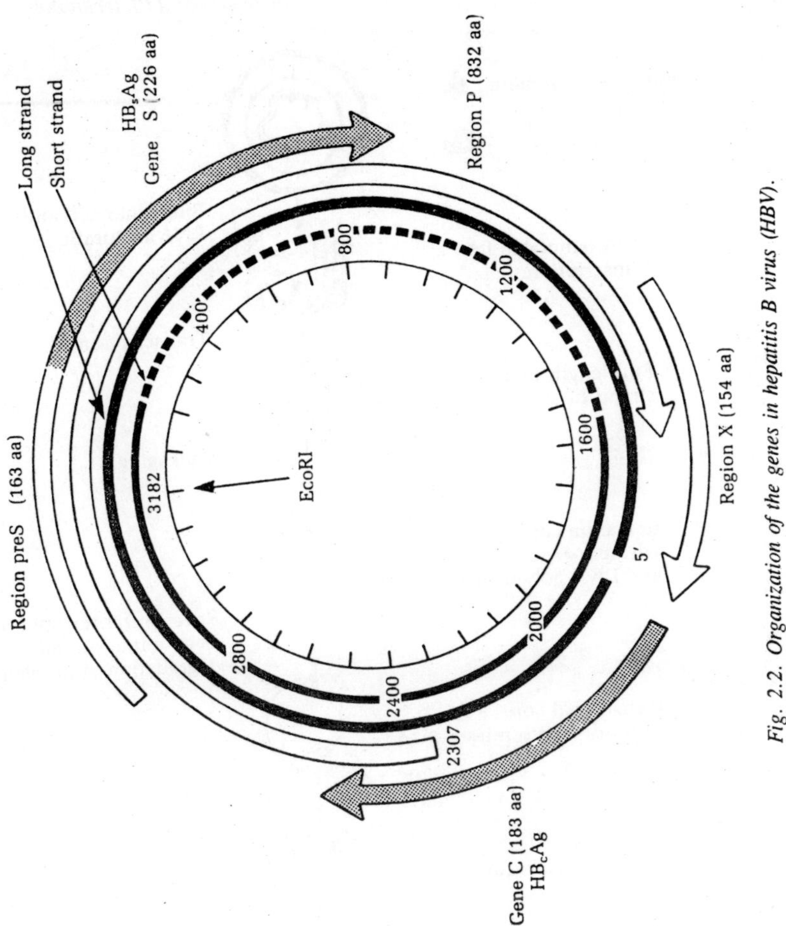

Fig. 2.2. Organization of the genes in hepatitis B virus (HBV).

Practical Utility of Viruses

Besides the usage of viruses in development on different vaccines, it is also used for many purposes.

1. *Bioinsecticide* : *Baculovirus* group is used as biocontrol agent against agricultural insects.
2. *Phage-typing* : Number of bacteriophages is used for *lysis* (killing) of bacterial cells to find particular strain of a bacterial species.
3. *Genetic Engineering* : Different viruses namely, M13 and phage in *E. coli* bacteria are commonly used as vectors to transfer specific genetic elements (DNA) from one organism to other through the infective pathway.

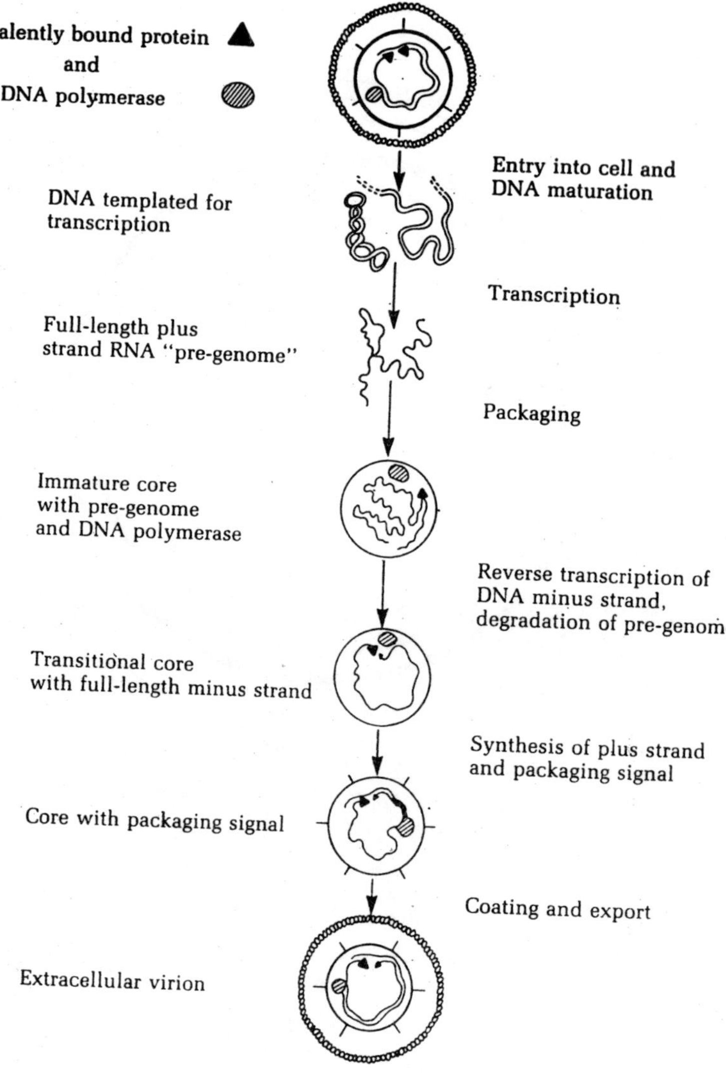

Fig. 2.3. A model for the replication of hepatitis B-like viruses.

Mycoplasmas (PPLO)

There are minute living cellular organisms, with no cell wall (unlike bacteria) and with loose nucleus, whose size starts from 100-150 mm. They are midway between virus and bacteria and were first noted by L Pasteur in cattle. They most time seen as spheroid, but change shapes (*pleomorphic*). In a growth medium, mycoplasma area

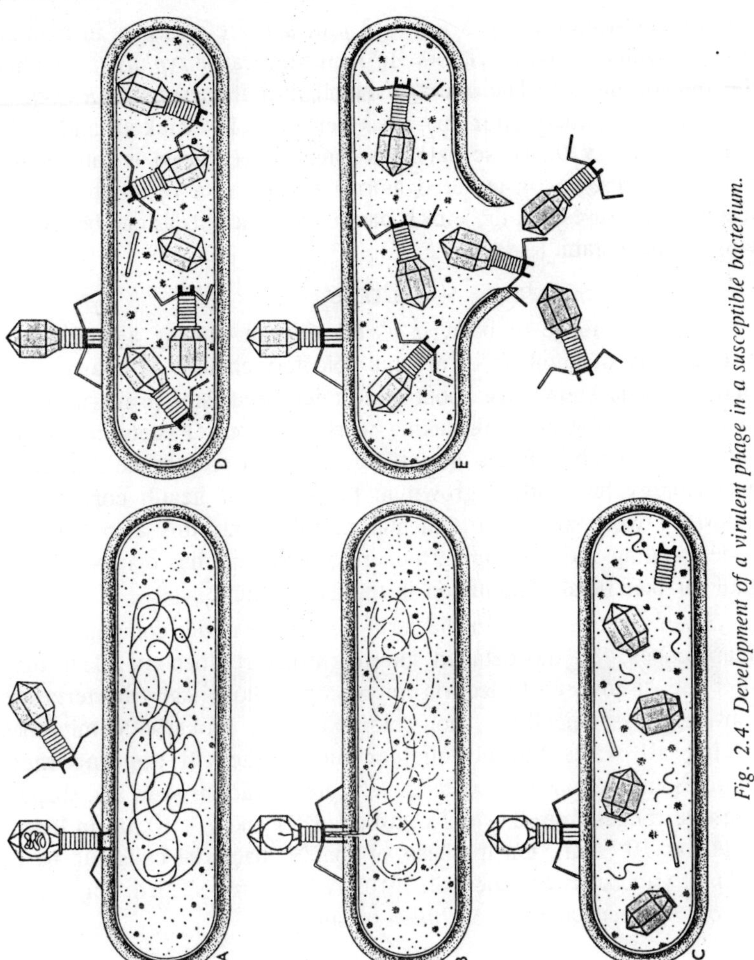

Fig. 2.4. Development of a virulent phage in a susceptible bacterium.

looks like a 'fried eggs', unlike viruses, they can grow on culture media. They are parasites in animals, plants and man. In cattle, *M. mycoides* cause pleuro-pneumonia (pleuropneumonia like organism or PPLO). It can cause respiratory infection (*M. hominis*), other species may produce arthritic disease, and can also affect central nervous systems (CNS). They reproduce by budding as do the yeast cells, and not by fission as happens in bacteria.

RICKETTSIAS

These are rod-shaped and pleomorphic bacteria like organisms with cell walls, which cannot move and reproduce by binary fission. Their size ranges from 0.2 – 0.5 μm × 2 μm, and are parasites in man an insects, which serve as victors. In man they cause *typhus fever* after the bite of insects. The average member of the group is *Rickettsia prowazekii*, named after the discoverers A.H. Rickett and C.S. Prowazek in 1910. These parasites cannot be cultured in lab media, but can be grown on cultured animal tissue, as a result they show features of viruses and those of bacteria also. Rickettsias can be stained with *Geimsa* stain.

BACTERIA

The knowledge of bacteria as the disease causing agents which has capacity of numbers of organic acid fermentation is known. Our first scientific knowledge about these microorganisms primarily owes to the pioneering and painstaking works of Louis Pasteur in France and Robert Koch in England in 19th century. At present the science of bacteriology has widely grown in the fields of health core, many industrial work and environmental management, this also helps in understanding and developing modern biology and has opened a large gate for beneficial exploitation in the 21st century.

Even when majority of bacteria can be seen with compound light microscopes, yet the detailed internal structures become clear only after the discovery of electron microscope. Biologically bacteria are considered as unicellular, non-chlorophyllous, prokaryotic microbes having all walls but no nuclear membrane surrounding their chromosomes. Shapewise bacterial cells are of various sizes and shapes. The size ranges from 0.5 to 1.0 μm diameter and 2 – 10 μm in length (1 μm = 10^{-3} mm). On the basis of shapes, bacterial cells can be of 4 types, (i) *Coccus* or spherical, (ii) *Bacillus* or rod-shaped, (iii) *Vibrio* or coma-shaped, and (iv) *Spirulum* or spiral.

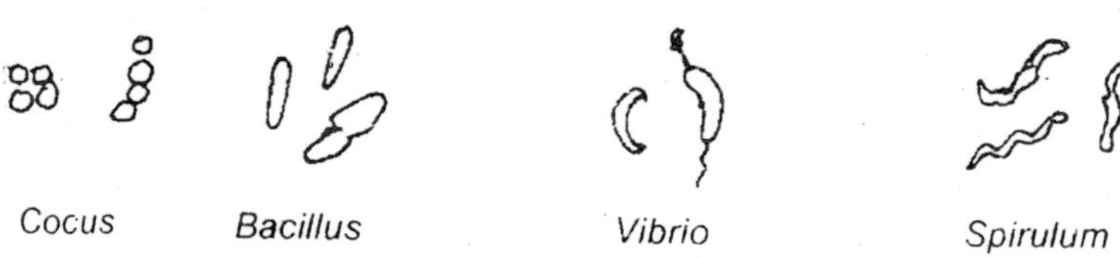

Fig. 2.5. Bacterial cell types.

Even though majority of the bacteria maintain a definite shape due to the presence of a harsh wall, some may be *pleomorphic* i.e., exhibit a number of shapes. Bacterial cells may remain in clusters, chains or filaments. They cannot move or swim around, while some may have flow movement on a solid surface. The motion forms have

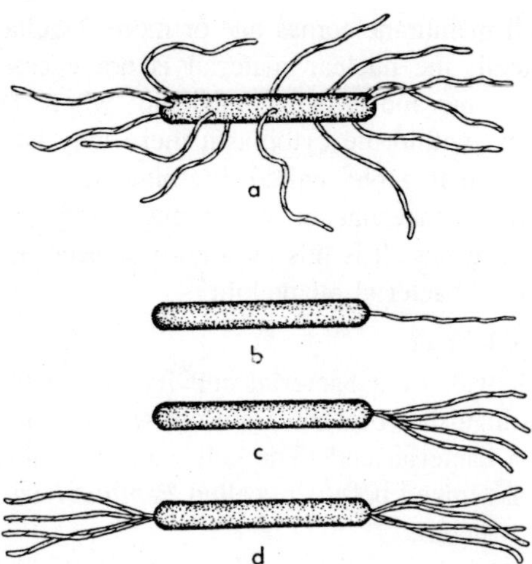

Fig. 2.6. Types of flagellation: (a) peritrichate; (b) monotrichate; (c) lophotrichate; (d) amphitrichate.

one or more flagella. Cells are classified on the basis of number of types of distribution of hairs/flagella.

1. Monotrichous
2. Bitrichous
3. Lophotrichous
4. Amphitrichous
5. Peritrichous

Besides from the wall and membrane, a bacterial cell may be covered by a *slime layer*, which when hardens, is termed as *capsule*.

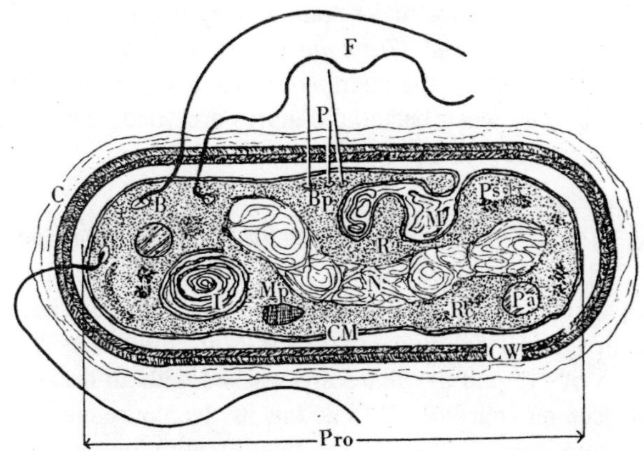

Fig. 2.7. Diagram of a typical bacterial cell showing all recognized structures.

From the cell membrane comes one or more flagella in motile species. Within the cell, the nuclear material is not encased with a definite nuclear membrane and is called nucleoid, where DNA is in circular form. Besides, within the cytoplasm there may be one or more very small free circular DNA called *Plasmid.* It does not directly take part in bacterial inheritance, yet it may carry some environmental stress-resistant genes. It is this plasmid now become petition in genetic manipulation of bacterial adaptability.

Bacterial Cell Wall

The cell wall of a bacterial cell has a specific significance. Its presence distinguishes bacteria from viruses and mycoplasmas. It gives stiffness to a bacterial cell. The cell wall is chemically composed of a porous cross-linked polymer, called *Peptidoglycan* or *Murein* and is of complex structure.

The nature of this peptidoglycan differentiates 2 broad groups of bacteria and their staining behaviour. In *Gram-positive* bacteria, this layer is thicker (5 nm) that of *Gram-negative* ones (2 nm). In *Gram-negative*, there is a thick outer membrane beyond the peptidoglycan which is rich in lipids and serves as barrier to external chemicals. *Gram positive* fails to have outer membrane bacteria and so their walls can easily be dissolved by the *lysozymes*. This peptidoglycan is absent in *Archaebacteria* and the wall is made of glycoprotein as in *Methanobacterium*.

Staining of Bacterial Cells

The usage of staining schedule for colouring bacterial cells is known as *Gram staining*, was developed in 1884 by Hans Christian Gram in Denmark is used as major stain to vary two groups of bacteria, namely, *Gram positive* and *Gram negative*, it depends on chemical nature of wall and retains the staining colour (violet) and serves as a differential stain only when bacterial cultures are young and mechanism of cell wall is undiscovered.

Staining procedure

The cells are fixed to glass slide by applying heat and stained with a basic dye, *crystal violet* (4% for 1/2 hr) which is firstly taken by all types of bacteria in similar quantity, and slides are treated (5 – 10 min) with I_2 – KI solution mixture (Grams Iodine) as a biting to help retention of dye and the slides are washed with an organic solvent mixture of acetone, ethanol (1:1) acting as decolourising agent. The slides are again counter stained (1 minute) with red dye *saffranin*, washed in water, dry and examined under microscope. Since Gram positive bacteria has peptidoglycan retains the original *violet* stain etc., *Bacillus*, *Clostridium*, *Lactobacillus*, *Streptococcus*, *Staphylococcus*, etc.

Whereas in Gram negative bacteria, with peptidoglycan layer, the violet colour is removed by acetone, ethanol mixture, hence they show the red colour of *saffranin* stain, etc., *Escherichia*, *Salmonella*, *Rhizobium*, *Pseudomonas* and *Enterobacter*.

Some features of gram-positive and gram-negative bacteria

The 2 groups are distinguished from one and another. Apart from gram stain, there are some stains for certain members. *Acid fast* staining is special feature of *Norcardioform* bacteria, namely *Norcardia*, *Mycobacterium* and *Corynebacterium*. This was formerly used by Paul Erlich in 1882 and since then acquired clinical important for fast identity of pathogenic genera. There is also negative staining with *India ink* or *Nigrosine* which stain the background and not bacterial cell, and also helps in study of features of bacterial cell, and it can also be studied with no staining with the help of dark-field and phase-contrast microscopes. There are some stains for *endospores* with malachite green and saffranin, while for *flagella* Ziehl's carbolfuchsin and for *capsule* Congo red, acid-alcohol and acid-fuchsin stains are used.

Methods of Identification of Bacteria

Since bacteria are small in shape, size and structure which is why a number of technique including growth and cultural behaviour, staining, biochemical reactions, immunotechniques etc. are restored to give a better ending.

1. *Staining* : Gram stain and other kinds are discussed.
2. *Cultural Characteristics* : Include form, elevation, margin and morphlogy of colonies.
3. *Fermentation of Sugar/glucose*, the capacity is noted and indicator is used to note the changes in pH and production of gas.
4. *Enzyme activity* : Enzymes like catalase, oxidase, etc., are produced or not.
5. *Serological activity* : Since bacteria can serve as an antigen, gram-positive is a poor antigen when compared to gram-negative species. The new techniques included the use of fluorescent dye FITC that is present with patients antibody, *Streptococci* will fluoresce in such a situation.
6. *Bacteriophage typing* : Some phages attack particular strain of bacterium. In those cultures where the bacteriophage attacks and kills the *lysis*, virus plagues are produced, helps to identify a strain.

Key to identify some bacteria :

I. Cocci

A. Gram-positive

(a) The enzyme catalase is produced ..../.... *Micrococcus lutens*, *Staphylococcus aureus*

(b) Catalase is not produced ..../.... *Streptococcus*

B. Gram-nagative

(a) Glucose fermented .... *Neisseria fucosa*

(b) Glucose not fermented .... *N. flavescens*.

II. Rods/Bacillio

A. Gram-positive

(a) Endospore present .... *Bacillus*

(b) No endoscpores .... *Lactobacillus*

B. Gram-nagative

(a) Glucose fermented .... *Enterobacter aerogens*, *Escherichia coli*

(b) Glucose not fermented .... *Pseudomonas*, *Alcaligenes*

Reproduction in Bacteria

Fission

Bacteria reproduce millions of cells within no time. The main mode is by *binary fission* i.e., division into 2 and due to this bacteria are also called *Schizomycetes*.

Spore formation

In some groups, minute hard spores are produced with a cell, the whole cell content compresses and later spores are released by cell decay and are called *endospores* which are resistance to drugs or other chemicals agents due to presence of thick double layered wall made of keratin like protein offering withstand and is a mode of sexual reproduction.

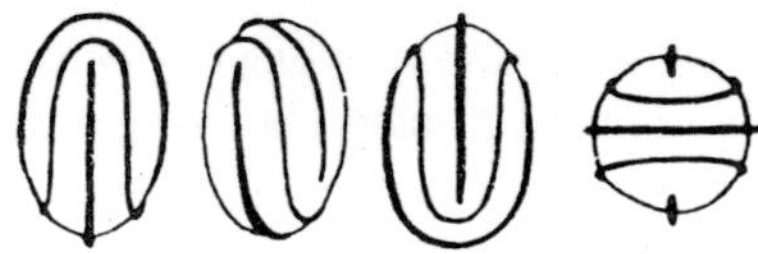

Fig. 2.8. Surface structure of a spore of Bacillus polymyxa. From left to right: side viewl same rotated a quarter turn from right to left; same rotated a further quarter turn; view of a pole.

Sexuality and genetic recombination

The purpose sex is to transfer genetic material from one to another and form a new combination in offspring, may be effected in bacteria in three ways.

Conjugation

In bacteria, sex was not known since long and in 1950s the works of Lederberg and Tatum and of Wollman and Jacob, it became clear that there is a system of mating for genetic transfer, even though the two sexes could not be morphologically featured. The tube like *sex pilus* developed with two cells of opposite 'sexes', through this tube the genetic material is transferred to the other, the donor may be considered as male (+) and the receiver as female (–). The male cell has a small circular DNA in cytoplasm called *sex factor* or fertility factor (F factor) is responsible for genetic union.

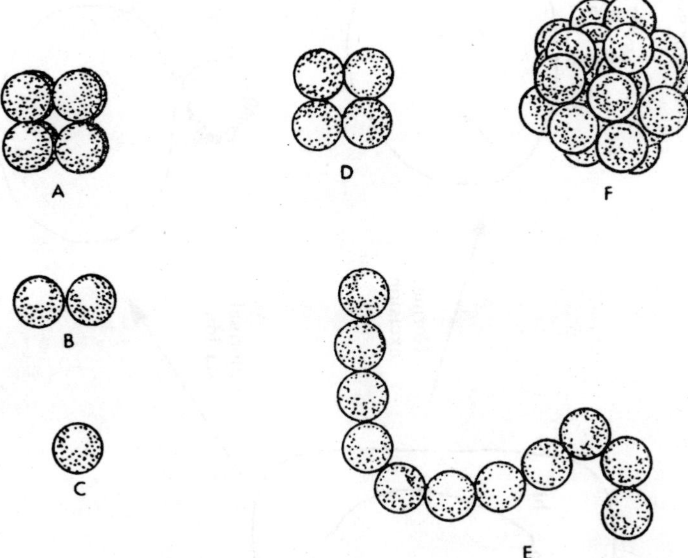

Fig. 2.9. Arrangement of cocci showing: A, fission in three planes at 90 degrees, forming cubical packets, e.g., Sarcina; B, fission in one plane, forming diplococci, e.g., Diplococcus pneumoniae; C, a single coccus, basis of all the arrangements and frequently seen in any of them; D, fission in two planes at 90 degrees, forming tetrads, e.g., Gaffkya tetragena; E, fission in one plane with adherence of the cocci, forming chains of streptococci, e.g., Streptococcus pyogenes; F, fission in various planes, forming irregular groupings or clusters, e.g., Staphyloccus.

Transduction

This is another mode of genetic transfer in bacteria considered by viruses (*bacteriophage*). Here a virus infects a bacterium and carries a bit of its genetic material after the bacterial death and is transferred to another after infecting it.

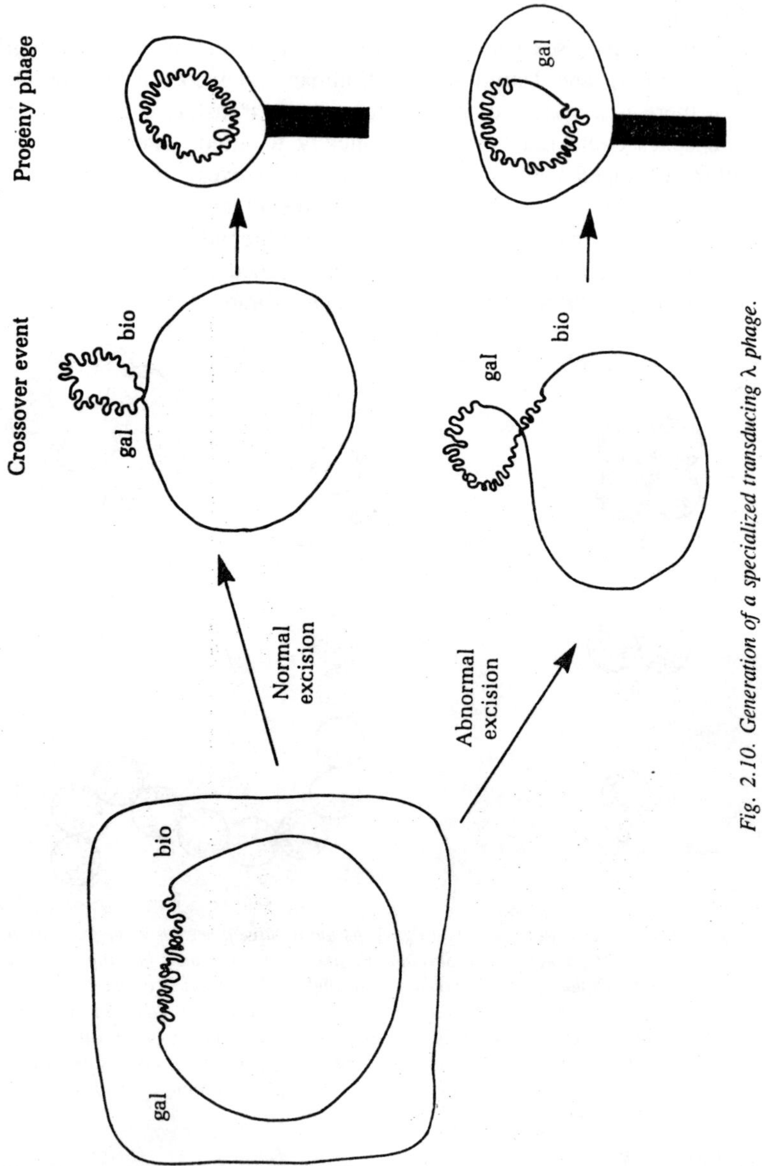

Fig. 2.10. Generation of a specialized transducing λ phage.

Transformation

Cell free genetic material gets transferred from one to another cell, which may occur in laboratory or nature conditions, and was noted by Griffith in 1928 and later, by Avery, Macleod and McCarty in the bacterial genus *Pneumococcus*.

Nutrition and Metabolism in Bacteria

Depending on the mode of nutrition, bacteria can be grouped as *autotrophs* or *heterotrophs*, majority of, which are heterotrophs, which are either *parasites* or *saprophytes*. Parasites either *obligate* ones who depend entirely on living host for organic food, or they may be powerful who sometimes get food from non-living organic source. On the other hand, Saprophytes get nutrition from decaying organic matter by putrefaction and fermentation. The usual saprophytic bacteria responsible for various types of separation of elements in nature are rather mapping in their nutritional requirements and get nourishment from number of organic substances, namely, carbohydrates, proteins, fats and amino acids and for types of metabolism they are supplied with a number of *extracellar* and *intracellular* enzymes.

For *autotrophic* nutrition, bacteria are of two kinds *Photoautotrophs* which can derive energy through process of photosynthesis, eg., green and purple bacteria with a special type of chlorophyll called *bacteriochlorophyll* and are energy like higher plants, whereas the other group is *Chemoautotrophs* which get energy for synthesis of organic compounds from oxidation of some other chemical compound (*chemosynthesis*), and in the following works are done by various bacteria:

1. Sulphur Bacteria (*Beggiatoa*) $H_2S \rightarrow S + 2H^+ + 2e^-$. Here energy is got through the break-down to hydrogen sulphide.
2. Iron-bacteria. The energy is arrived from iron compounds, where ferrous iron is converted to ferric iron, eg. *Leptothrix* $Fe^{++} \rightarrow Fe^{+++}$
3. Hydrogen Bacteria. They oxidize molecular hydrogen to form water with the release of energy (purple bacteria).
4. Nutrifying Bacteria. $NH_4^+ \rightarrow NO_2^-$ (*Nitrosomonas*) $NO_2^- \rightarrow NO_3^+$ (*Nitrobacter*)

The ammonia is converted to nitrite and the latter species then oxidizes into nitrate. The metabolic processes in bacteria are linked to mode of respiration and based on these conditions, bacteria are divided into

1. *Aerobes* which required free oxygen
2. *Anaerobes* thriving in conditions devoid of oxygen where the metabolish (*fermentation*) is solely under the control of certain enzymes.

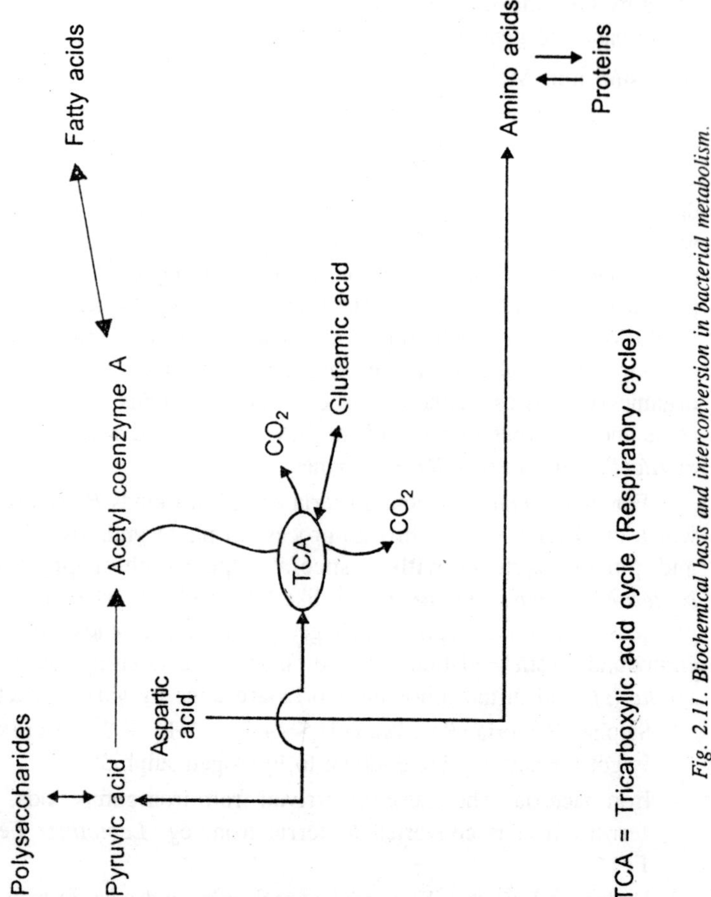

Fig. 2.11. Biochemical basis and interconversion in bacterial metabolism.

3. *Facultative anaerobes*, the organisms can metabolize according to the presence or absence of oxygen, so that the oxidation products differs, i.e., either CO_2 and H_2O or CO_2 and some intermediates such as alcohol. Due to low heat production anaerobes are preferred in organic sludge treatment so that huge amount could be treated.

Bacterial growth and decay

This refers to increase in number of cells in a given time, which is more among the microbes under suitable nutritional and environmental conditions and is being effected by division of bacterial cells and the increase in population increases in geometrical progression i.e.,

$$1 \rightarrow 2 \rightarrow 2^2 \rightarrow 2^3 \rightarrow 2^n$$

where *n* refers to the number of generations covered during the period. Growth can be quantified either by cell number or by cell mass.

In terms of number, the number of generation (*n*) can be calculated as:

$$n = \frac{\log_{10} N - \log_{10} N_0}{\log_{10} 2},$$

where *N* and N_0 are populations at the end and beginning of a given time.

Now $\log_{10} 2$ is known as 0.301, therefore,

$$n = \frac{\log_{10} N - \log_{10} N_0}{0.301}$$

$$= 3.3(\log_{10} N - \log_{10} N_0)$$

While *generation time* or *doubling time* (g) says the number of generations completed within a given time *t* when the original population becomes double, Thus

$$g = \frac{t}{n}, \text{ and the rate of growth } r = \frac{n}{t}$$

To quantify g has been done for certain bacteria, under a particular temperature (37^{0}C) condition, which is listed below :

Organism	g (min)
E. coli	12.5
Staphylococcus aureus	27.30
Mycobacterium tuberculosis	792
Pseudomonas putida	45
Lactobacellus acidophilus	66

Phases of growth

The growth of population follows four phases when a sample bacteria is transferred from one to another and the growth curve takes a S-form and then declines.

Log phase

When population get a new climate at initial stage is followed by active growth phase follows rapid divisions called *log phase* or *exponential phase*. Then growth gradually comes down either by the use-up of nutrients of toxic metabolic products or some other growth limited factors and finally stops and this period is static and termed *stationary phase*, and lead the cells die and leads to loss of capacity to reproduce, however death rate differ with the organisms and the

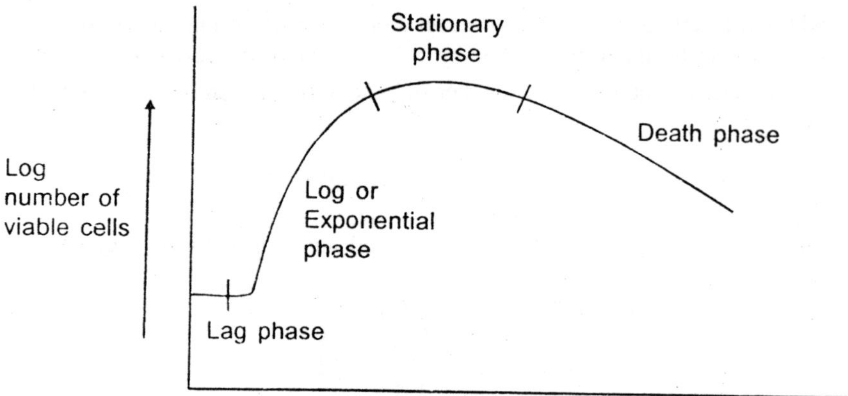

Fig. 2.12. Bacterial growth curve.

environmental factors. Some bacteria die and consequently few viable cells remain after 72 hours eg., *Bacillus* species, while enteric species die very slowly and may continue for months.

Measurement of growth

Bacterial growth can be quantified either by cell mass or cell number.

Cell mass

Cell mass can be done directly weighing the biomass through the determination of turbidity by optical means. The turbidity in water sample due to bacterial growth is got through turbidometer where photoelectric measurement in based on the decrease in light transmission due to increase in assumed that increase in OD is related to increase in cell number, other words, OD is directly proportional to bacterial

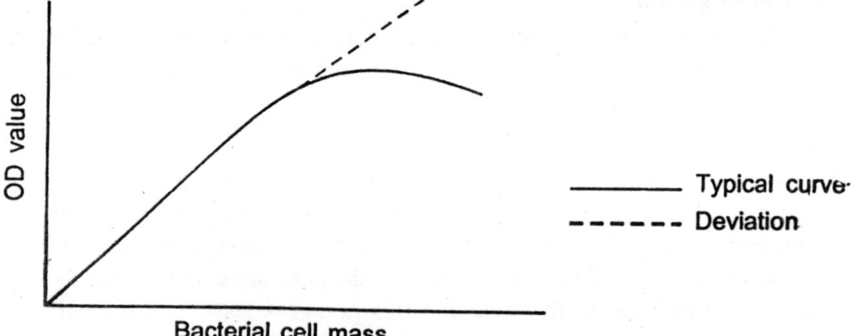

Fig. 2.13. Direct proportionality of biomass with OD.

cell density, otherwise transmission of light is affected and give wrong data.

Cell number

Bacterial cell is measured by a particular amount of water sample by diluting it and directly count cells either under a microscope or through electrical particle counter (*Coulter counter*) which finds the presence of particle causing electrical in store and displays on a digital reading where water should be free of any other debris. Cell estimation can be done by culturing a drop bacterial suspension in agar plates hatched for a fixed time and at fixed temperature and finally count the number of colonies developed on plate (*colony count*), wherein each cell is supposed to grow into a colony after a certain period. The number of colonies is multiplied by the dilution factor. Estimation could include total number of cells living or dead, which accounts for living cells only.

Factors affecting bacterial growth

(i) *Nutrient availability*. Growth of bacterial population is influenced by type available nutrients like vitamins, aminoacids etc., and their proper way through the cell membrane. Nutrient which deficient mutated in a bacterial species, where particular nutrient has to be supplemented for proper growth.

(ii) *Temperature*. Of temperature requirements, bacteria have been divided into 3 groups, (a) *Thermophiles* which can grow at temperature above 40°C–45°C white some grow in hot springs, (b) *Mesophiles* normally grow in the temperature range of 25°C–40°C though the optimum temperature is 37°C for most microbes, and (c) *Psychrophiles* actively grow below 20°C and are cold tolerant and often found in arctic and antartic regions. *Thermal resistance* of bacterial spore also be calculated, bacterial spores are more withstandable to temperature than vegetative cells, where D_{250} is the time in minutes referred to *one log cycle* (90%) reduction in population of a given microbe in a given temperature of 250°F.

For *clostridium* B, D_{250} = 0.21 minute

From 1 spore to 10^{-12} spores = 12 log cycles

Thus if 1 log cycle = 0.21 minutes at 250°F, then for 12 log cycle, time required = 0.21 x 12 = 2.52 mts. At 250°F

For vegetative cells it taken 1 minute at 160°F.

(iii) *pH*. Influences microbial growth by action on enzymes work and cell permeability. Normal growth can occur at a pH range of 6-

8, when few can bare below 4, eg., for *E. coli* it is 4-7; for *Thiobacillus*, 1-4, while for nitrite bacteria, it ranges from 6.5-9.

(iv) *Osmotic pressure*. At higher osmotic potential, microbes take water by osmosis while at lower OP, water come out of cells (*exosmosis*), this is exploited in food preservation with higher salt use. *Halophiles* are some species, which can grow at very low osmotic potential.

Control of bacterial growth

This can be controlled either by physical or chemical means. When a germicidal agent kills a bacterium by way of lysis of cell membrane called *bacteriocidal*. Otherwise, if it only prevents multiplication is called *bacteriostatic*.

Physical agents

This includes temperature UV-radiation, ultrasonic waves, etc., temperature above 120^{0}C or below 20^{0}C is detrimental. Thermal death time refers to the shortest period to kill bacterial cells in suspension. For *E. coli*, it is 20-30 mins, at 57^{0}C, whereas for *Salmonella* it is 4 mins at 60^{0}C.

UV-radiation. The germicidal effectiveness of ultraviolet radiation lies at 2600 Å which splits the protoplasm and enzyme system.

Ultrasonic waves. Vibration of ultrasound (>20000 H_2) causes cavity in water around its microbial cell and thus puts a mechanical pressure causing separation of protein and splitting of the membrane.

Chemical agents

Some of the chemical compounds have the ability to inhibit bacterial growth apart from antibiotic drugs where some are phenol

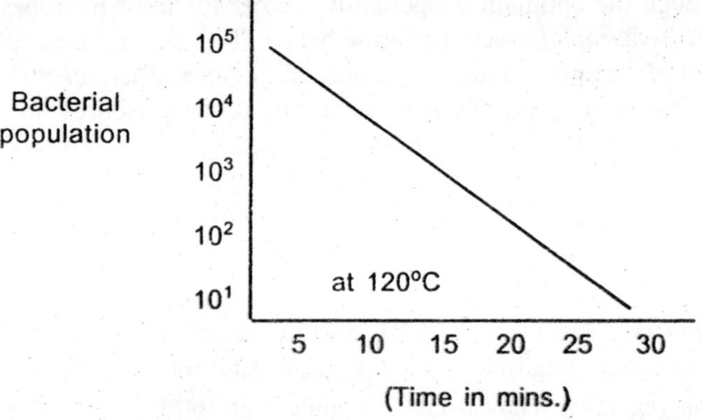

Fig. 2.14. Thermal death curve.

and its derived, acids and alkalies, salts like NaCl which change the cellular ionic balance, detergents, chlorine, heavy metals and organic compounds like PAB and PAS. Phenol is a bacteriocidal compound, which kills bacteria by damaging the membrane and dispossess the protein and was first used by Joseph Lister in 1880.

The effectivity of a chemical agent is got by a comparative test with reference to phenol and is done by determining the *phenol coefficient* of compound, which is the ratio of minimal sterilizing concentration of phenol to that of compound. A germicide is generally recommended for use at 5 times this concentration.

$$\text{P.C.} = \frac{\text{Reciprocal of the dilution of the agent}}{\text{Reciprocal of the dilution of the phenol}} = \frac{X}{P}$$

Potassium permanganate is not a very powerful disinfectant, unlike chlorine, but HgCl at 1 ppm can control microbial growth. Microbial antagonism may also deter the growth among different microbes, which is effected through the production of toxins. The toxin *colicin* produced by plasmid of *E. coli* in the gut of man, can control the development of *Salmonella* and *Shigella*.

Bacteriology of Water and Sewage

Bacteria play an vital role in acquatic system by decomposition of organic matter. Microbes are encouraged by organic contents due to richness of water, of which some could be pathogenic and cause various diseases in both man and animal of which some are cholera, typhoid, paratyphoid, dysentery, bacillary dysentery, gastroenteritis, etc., which may turn into epidemics and cause damage. The organisms responsible are *Bacillus*, *Vibrio*, *Salmonella*, *Klebsiella*, *Shigella*, *Clostridium*, *Strepotococcus* and *Esherichia*.

Microbial flora in potable water primarily originates from untreated or partially treated sewage, tannaries and food processing industries. All of them are pathogenic and some pathogenic ones do not survive long, yet we have so many water borne diseases that sometimes turn into epidemics. The degree of microbial pollution varies with type of natural water which is primarily classified (1) *atmospheric water* from precipitation (2) *surface water* derived from rain and run-off (3) *stored water* in tanks and pools (4) *ground water* much below the upper soil layers, of them ground water is least polluted by bacteria due to the filtering action of soil layers and the absence of organic matter. Drinking water must be free from taste, odour and toxic chemicals apart from pathogenic organisms.

Bacteriological Examination of Water

This is done periodically to find the pollution from domestic sewage which contains lots of pathogenic and non-pathogenic forms and to cut their possibility of transmission.

Often the examination is based on non-pathogenic forms, that too, not on a single type, but from a group of them commonly called *Coliform*, and whether they are present or absent in the test water sample. It can be measured and is done from the approximate numerical determination MPN (Most Probable Number).

Indicator or Index Organism

The most important groups of non-pathogenic enteric bacteria is sewage polluted water are *Escherichia coli* (rod-shaped) and faecal streptrococci, *Streptrococcus faecales* (round cells) and are selected as indicators due to their constant presence in such type of polluted water alongwith the pathogenic ones and themselves are non-pathogenic. Others are *Clostridium perfingens* and *Biefidobacterium bifidus*. They are identified by simple cultivation methods.

Why Coliform?

Detection and estimation of pathogenic bacteria in water is boredom, time consuming and laborious whereas the *coliform* groups is easily tested and enumerated, which indicates that the pollution is of faecal origin and accordingly, suitable steps can be taken within a short period. Members of this group remain alive for a longer time and allow sufficient time for testing. But faecal streptococci survive for lesser time that *E. coli* so their presence in high number in polluted water indicates that the pollution is of recent origin.

The advantage of such microbial tests help in distinguishing pollution of faecal origin from the non-faecal one. Raw sewage approximately 10^7 *E. coli*/100 ml. On the basis of bacterial estimations, some *microbial standards* of water quality from different sources have been formulated, namely,

Highly polluted	≥	10,000	bacteria/litre
Moderately polluted	≥	1,000	bacteria/litre
Slightly polluted	≥	100	bacteria/litre
Water of satisfactory quality	≥	10	bacteria/litre
Pure drinking eater	<	3	bacteria/litre

Routine Bacteriological Analysis

Proper sample collection is aseptic conditions, dechlorination with sodium thiosulphate and their testing in laboratories within shortest possible time is essential, although samples can be preserved at 4^{0}C for 6 hours. But in no case the sample can be kept for more than 24 hours to avoid rapid changes in bacterial content, methods of analysis comprise various tests that indicate the presence of coliforms and pathogens.

Standard plate count (SPC)

It is done to enumerate total viable population and not to detect either coliform or other pathogenic forms present therein and check the efficiency of water treatment methods, and is done periodically. A bacterial spore is assumed to form a colony. It is expressed in numbers per ml.

$$\text{SPC} = m \times 10^n$$

where m is the number of colonies and n is the dilution factor.

Tests for coliforms

There are two standard methods for estimation of total coliforms in water sample, namely multiple tube fermentation (MTF) test and membrane filter technique. Both techniques are in use and each of them is recognized as a reliable method.

Multiple tube fermentation

It is essentially an acid fermentation system, where fermentation of lactose broth is tested at 35-37^{0}C for 24 and 28 hrs. in a series of test tubes involves three sequential tests.

1. *Presumptive test*. The source sample is diluted many times as per the expected bacterial population and is allowed to incubate in the growth medium for 24 hrs. The acid and gas production is checked after the given period. If gas is formed, then test is positive, if incase of failure, the test is negative.

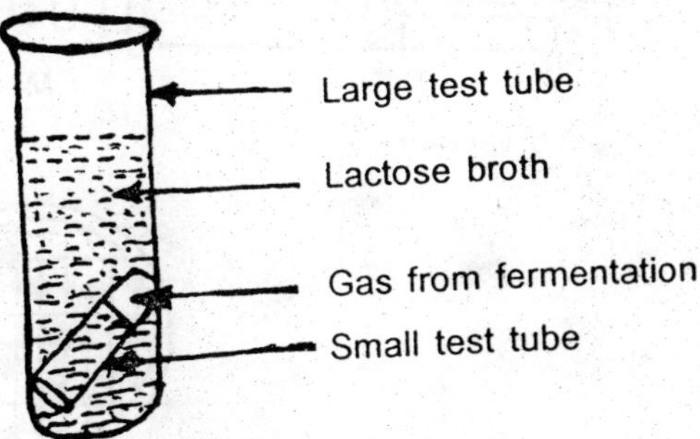

Fig. 2.15. MTF test.

2. *Confirmed test*. If there is a negative result, it is again incubated for another 24 hrs. if gas forms, it confirms the presence of coliform. On the other hand, if incase of no gas, even now, the test is considered negative, i.e., it can be taken that there is no coliform in the tested sample. It is a confirmed test for the presence of coliform, because gas production in previous test could be due to the presence of organisms other than coliform.
3. *Completed test*. This test ascertains the presence of coliform in the sample. Here a portion of sample from the positive test is grown on Endo or EMB (eosin and methylene blue) agar plates and is incubated for 24 hours at 35^{0}C. In this, E.coli will produce blue-black colonies with a greenish metallic sheen; whereas other members of the group will produce pale pink colonies without a metallic sheen.

Membrane filtration technique

This was first suggested by Goetz and Tsureishi in 1951 and named *molecular filter technique*. MF is made of cellulose acetate ester with a pore size of 0.3 to 0.5 mm. The filtration unit has the effective diameter of 10 nm. The membrane is first sterilized at 80^{0}C for 20 minutes in distilled water and then put on the filtration unit aseptically, is fixed to a vacuum pump if required, special type of syringe is also used for proper placement of the membrane.

When water sample is poured, the bacteria are held on the membrane surface. The filter with the trapped bacteria is dried and

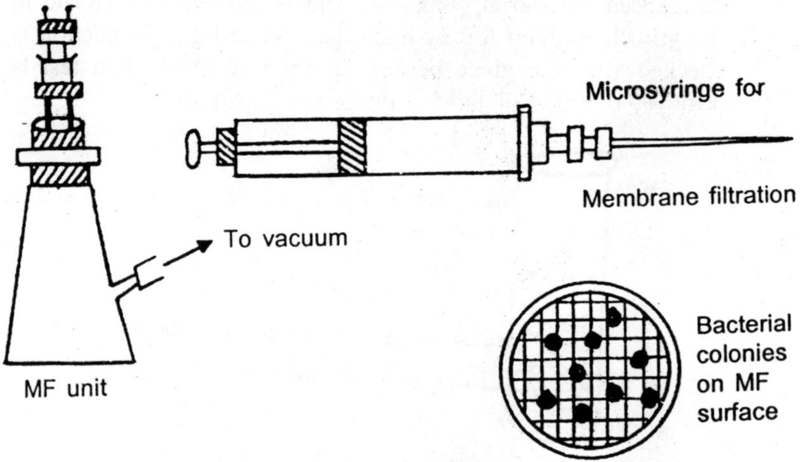

Fig. 2.16. Membrane filtration.

stained with a red dye called *erythrosine* for 30-60 mins. The filter is now placed on another filter saturated with distilled water and repeated to remove the excess stain, finally pink colour remains and filter is microscopically examined and the microbes counted.

The result is

$$X = K = \frac{N}{V}$$

where

X = total no. of microbes

V = volume of water (usually 1 ml)

N = average number of microbes in one microscopic field of view

K = Co-efficient of microscope, i.e., ration of the size of the field to the area of membrane filter.

Statistically significant results need large sampling. The advantage of MF over MTF is that it can be done on the field with use of hand pump which cannot be used in case the water is thick and heavily polluted, as the filter gets blocked. The degree of pollution is expressed in terms of (a) *Colititer* which is the smallest volume of water containing one *E. coli*; and (b) *Coli index* that is the number of *E. coli* present in 1000 ml. of water. However both MTF and MF methods are subject to false-positive and false-negative results. Hence a rapid *colilert system* has been developed for detection of total coliform and *E. coli* in water. This method at the same time can determine the presence of total coliform with yellow colour as indicator and production of fluorescence in same tube demonstrates the presence of *E. coli*, which is based on the concept that each target microbe, there is a substrate for a specific enzyme. When the microbe digests the substrate, a *chromogen* is produced. The substrate for total coliform in this case is (o-nitrophenyl) β-D–galactopyrannoside) which turns yellow, while *E. coli*, the substrate is (4-methylumbelliferyl β-D-glucuronide) produces fluorescence.

Methods for Differentiating Faecal from Non-faecal Coliforms

Out of enteric bacteria which can undergo chemical change lactose by producing acid and gas, all may not be completely intestinal parasites like *E. coli*, eg. *Enterobacter aerogenes*, though can undergo lactose, is also found in nature in decaying materials, therefore its presence in water may not indicate a faecal contamination. Thus it is necessary to determine whether the coliform is of faecal or non-faecal origin.

Differences between 2 organisms

	Characters	*E. coli*	*Ent. Aerogenes*
(i)	Colony size	Smaller (2-3 mm)	Larger (4-6 mm)
(ii)	Colony surface	With metallic sheen	Without metallic sheen
(iii)	Colony confluence	Discrete	Tend to coalesce
(iv)	Colony elevation	Concave	Convex
(v)	View under reflected light	Dark	Light brown
(vi)	Incubation at 44°C for 24 hrs.	It can grow	It cannot grow

However, to get a final difference between the 2 organisms and to conclude the faecal or non-faecal nature of pollution, a series of physiological/biochemical tests are done which are collectively known as IMViC test.

IMViC test

This consists of a series of four tests to identify *E. coli* and *Enterobacter aerogenes*. The letters in IMViC have different connotations which are as follows :

'I' — Represents Indole Test

'M' — is the methyl red test

'V' — stands for voges-proskauer test

'C' — is the citrate test, while 'I' is used only for purpose and has no significance with the test systems.

Indole test

This test is based on protein catabolism where indole is produced from tryptophan which indicates whether the enzyme tryptophanase is present or not in the test organism.

Tryptophan + O_2 ⟶ Indole + Pyruvic acid

(Kovac's Reagent used is p-dimenthyl aminobenzaldehyde) ↓ Kovac's reagent gives bright red colour indicating indole production)

Methyl red test

This test is for acid production from glucose and is actually a measure of pH when sugar is broken to acid, and at pH lower than 4.5, the red colour produced is a sufficient indicator of acid formation.

Voges-Proskauer test

It is a colour test of acetoin production from glucose, which is an intermediate stage in the production of butanediol from pyruvic acid.

Acetyl Methyl Carbinol (AMC) is presence of 40% KOH and air is further oxidized and gives the characteristic pink colour within 2-4 hrs, which is a positive test for *Enterobacter aerogenes*.

Citrate test

Attempts to get the ability to use citrate from Sodium Citrate as the sole carbon source *E. coli* has got the requisite enzyme citrate permease and cannot use the same, whereas *Ent. Aerogenes* can assimilate it and grow so that turbidity develops during the culture.

IMViC test differences:

Organism	*Tests*				
	I	*M*	*V*	*C*	*Pollution source*
E. coli	+	+	–	–	Faecal
Ent. Aerogenes	–	–	+	+	Non-faecal

Most Probable Number Estimation (MPN)

It is the statistical expression of estimating the number of bacterial cells in a culture, which gives an approximate estimation of the number of coliform bacteria present per ml. in a given decimal serial dilution in the MTF test.

The value is got by assuming that the smallest gas-forming portion in a geometric series has only one coliform organism. Therefore, reciprocal of volume expressed in ml. will equal the number of organisms per ml, which often happens that from a combination of positive and negative results where one has to drawn an approximate value. If the fermentation tube containing 0.1 ml of the same gives positive result, while the tube with 0.01 ml. does not, then the MPN is considered as 10 per ml. Statistical estimation is made and compared with MPN index value table (APHA). In fact, a sanitary bacteriologist

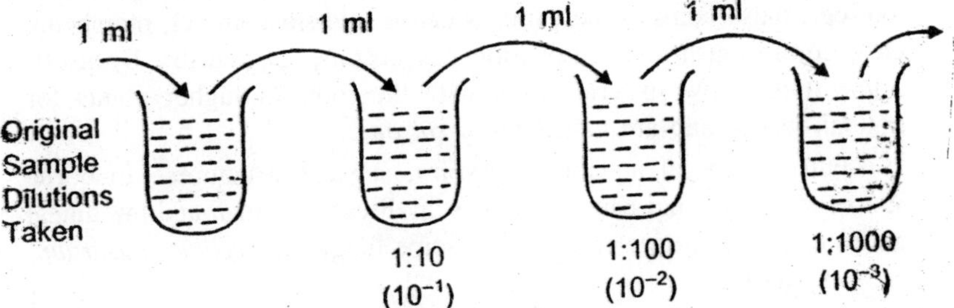

Fig. 2.17. Serial dilutions of sample.

is not specifically interested in exact number, but the approximate one to take effective control measures.

FUNGI

These are eukaryotic, spore-bearing non-chlorophyllous organisms and include molds and yeasts. Molds have multicellular filamentous bodies and yeasts are unicellular and have wide range of appearance from small microscopic filaments to large macroscopic structures. The filament is *hyphae*, which form a wooly mass called mycelium. These are made of several cells put end to end, or within one long thread may be of many nuclei, which is called *coenocytic*. The mycelia may unite and form a regular macroscopic vegetative structure called *fruit body* as in higher groups of fungi. The study of this group of organisms is called *mycology* (*mykes* = mushrooms)

Chief Characteristics

1. Lack of chlorophyll (green pigments)
2. Cell wall is made of *chitin*, instead of cellulose.
3. Food reserve is *glycogen* and not starch as in plants.
4. Nuclei within cells are very small.
5. Major mode of reproduction is a sexual by *spore* formation and some may reproduce sexually by conjugation.
6. They may be sub-terranean, terrestrial. Some grow as parasites on plants or on organic food stuff and make them rot. There are also pathogenic forms causing diseases in human beings, plants and animals.

Although fungi are economically important for some industries like pharmaceutical yielding drugs and breweries in the alcohol fermentation, cause heavy loss as parasites of crop plants and forest trees and in rotting food materials. Some of these fungi are edible too, eg., mushrooms of the varieties called toad-stool, morel, morchella, etc., edible mushroom cultivation has become an industry by itself. Some fungi grow in association with the roots of higher plants for mutual benefit and are called *mycorrhizae*.

Fungi grow in aquatic reservoirs, sewage structures, air-borne spores of many fungi cause various skin diseases of man and the fungal toxins called *aflatoxins* produced by some fungi (*Aspergillus*, *Fusarium*) are very harmful.

Classification

Fungi are broadly divided into 3 groups :

Lower fungi or phycomycetes

These are alga like filamentous, primitive and microscopic. The mycelium is non-septate i.e., *coenocytic*. No fruit body is produced. Spores are produced endogenously. Aquatic fungus *Saprolegnia* belongs to this having a long much-branched multinucleate mycelium. *Mucor* and *Rhizopus*, are called bread molds, and belong to this group. *Rhizopus arrhizus* mycelium is now used to recover minerals from mine waste effluent through adsorption on its biomass.

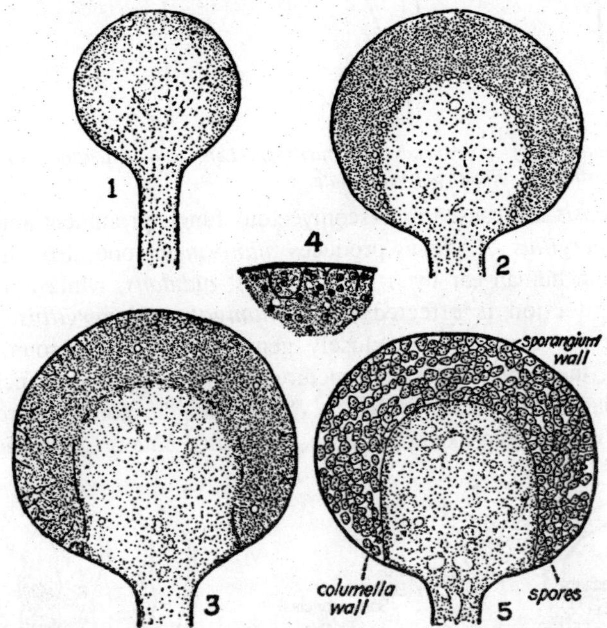

Fig. 2.18. Mucorales. Stages in formation of a sporangium filled with spores.

Higher fungi

These have definite bodies made of compact mycelia, which have septate hyphae with partitions formed in them, and are evolutionary advanced and are of 2 sub-groups

Ascomycetes or sacfungi

They are called due to the production of sexual spores endogenously within sac-like structures called *Ascus*, while a sexual spores (*Conidia*) are produced exogenously. They could be unicellular or multicellar filamentous, and certain species form distinct fruit bodies as in *Morchella* and *Tuber* or *Tuffel* and are edible. Many skin diseases or

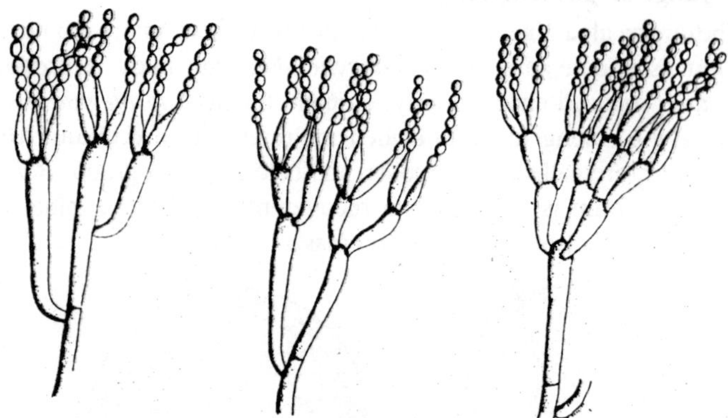

Fig. 2.19. Branching of spore heads in Penicillium. Left, monoverticillate; Center, biverticillate; Right, polyverticillate.

dermatomycosis are caused by Ascomycetous fungi, prominent among them in *Aspergillus*. *A. flavus* produces *aflatoxin* in food also causes otomycosis in human ear by *A. niger* and *A. nidulans*, while a T.B. like lung infection is effected by *A. fumigatus*. *Aspergillus* and *Penicillium* are common and closely genera within this group. In *Aspergillus*, the conidia bearing structure conidiophore is unbranched with a broad round head, while in *Penicillium* the conidiophore is branched with a broom like appearance, and called ***blue*** or ***green mold.***

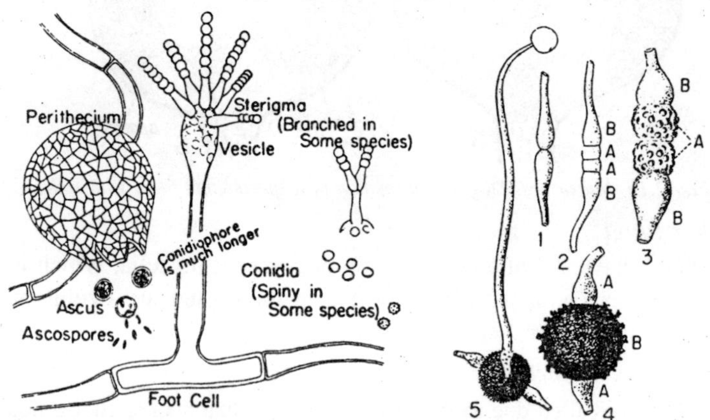

Fig. 2.20. Sexual spore formation. Left—Aspergillus; Right—Mucor.

As conidia are bluish green in colour and they grow on decaying fruits like oranges. *Penicillium* is medicinally important, being the source of the alkaloid drug *Penicillin* which is active against Gram positive bacteria. Other dermatomycoses include *tinecaptis* and *tinepedis*. The former is characterized by raised patches of scalp skin caused by *Mycosporium* species, while the latter is that of feet skin caused by *Trichophyton* species. The ergot-fungus *Claviceps*, on the other hand, is commercially important as it grows on rye plants and produces a hard black structure in the inflorescence of the plant, called *ergot*. Two important alkaloids are derived from these ergots namely, *ergometrine* and *ergotamine*, which can check haemorrhage during child birth, but when these ergots are consumed by cattle in the rye-crop field, abortion of calves can occur. The genera like *Neurospora* and *Sordaria* are well known, because lots of genetic researches have been done on their mutant forms in 1950s and 1960s yielding some classical theories. The other important member of this group is the unique unicellular fungus called *yeast*. Common genera are *Saccharomyces* and *Candida*, which help in the fermentation of alcohol from sugar and mollases. Most important species is *Saccharomyces cerevisae* or brewer's yeast used in beer industry. On the other hand, they are rich in vitamins and proteins, and are often mixed with other food stuff. They are minute (8 μm $\times$ 4-5 μm) spherical or oval in shape and mostly reproduce by fission. Therefore, are called *bud fungi* and also reproduce sexually by conjugation at some stage of life cycle. A few species are parasites on plants, while some are pathogenic to man attacking central nervous system, skin, liver and spleen.

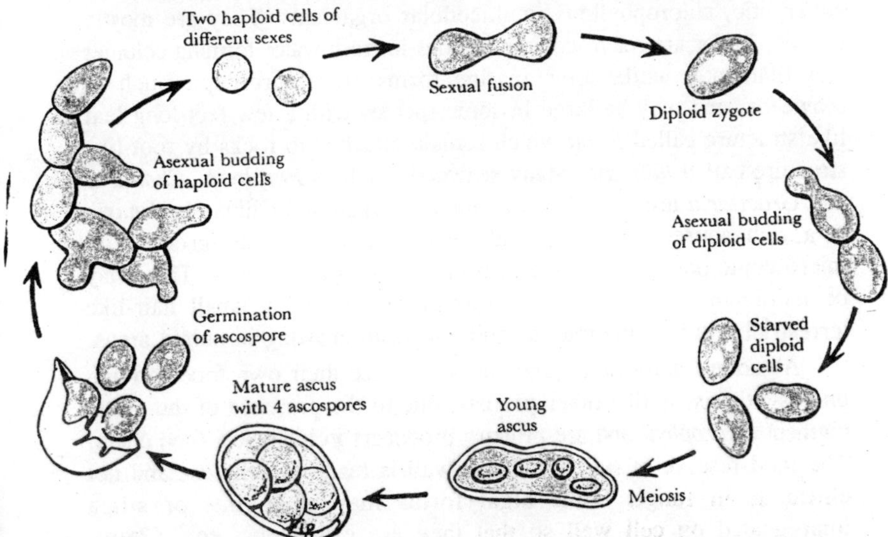

Fig. 2.21. The life cycle of a yeast.

Basidiomycetes or club-fungi

This groups is characterized by the formation of 4 exogenous sexual spores on the tip of a club-like structure called *basidium* and spores called *basidiospores*. This group contains many edible forms commonly known as toad-stool or mushroom including *Agaricus*, *Boletus*, *Volvularia*, *Calocybe*, etc. But some of them may be highly poisonous like *Amanita*, so one can ought to be careful while choosing the edible species.

Many of the basidiomycetous fungi grow on logs of wood and make them rot. Many of them look like brackets and are called *bracket fungi*. However, some of these fungal fruit-bodies such as *Polyporus*, *Ganoderma* and others are used as adsorbents of pollutants from water. Although basidionycetous fungi are in general, not microscopic, yet due to the aforementioned reasons, they are included here.

Deuteromycetes or imperfect fungi

These are microscopic, and life cycle is not perfect yet the name. They cause number of plant diseases, so they do not come under the purview of present discussion. Examples of this type of fungi are *Colletotricum*, *Fusarium*, etc. However, by and large all air borne fungal spores can affect the human body through inhalation.

ALGAE

This is a major division of *protists*, where most of species are eukaryotic, chlorophyllous, multicellular organisms. They are mostly aquatic. Major are microscopic planktons in fresh water forming colonies and filaments, while some marine forms are macroscopic such as seaweeds, and may be large in some species with a few feet long leaf like structure called *frond* which remain attached to rocks by root-like structure called *holdfast*. Many seaweeds such as *Porphyra*, *Chondrus* and *Gracilaria* are considered as food in Japan and China. *Laminaria* is a rich source of iodine and form a heterogeneous group with microscopic ones comparable to bacterial sizes and shapes. They may be motile or non-motile, with former group having small hair-like terrestrial, while some may be sub-terranean growing in humid areas.

Algae are *autotrophic* and can synthesize their own food through photosynthesis, unlike other protists, due to the presence of the green pigment *chlorophyll* and are primary producers in biological food chain. The food reserve in starch and cell wall is made of cellulose and not chitin as in fungi. Some algal forms may have lime or silica impregnated on cell wall so that they are calcareous, eg., *Chara*,

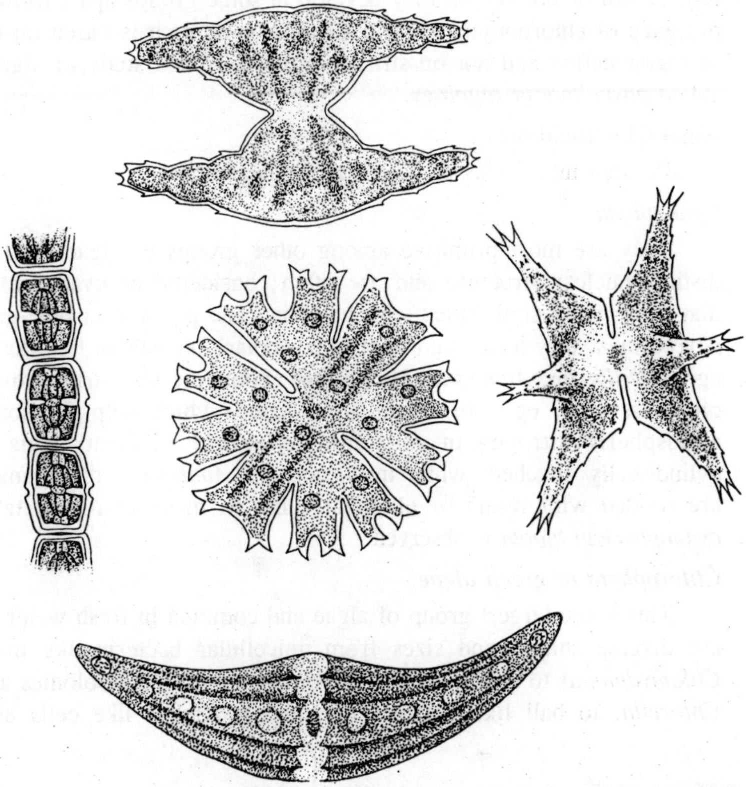

Fig. 2.22. Several different species of desmids, unicellular green algae, highly magnified, showing the symmetry of the cells.

Codium, *Diatoms* and *Halimeda*. They reproduce vegetatively by division, asexually by spore-formation and sexually by conjugation. Although the main pigment in algal cells is green, yet some other pigments such as

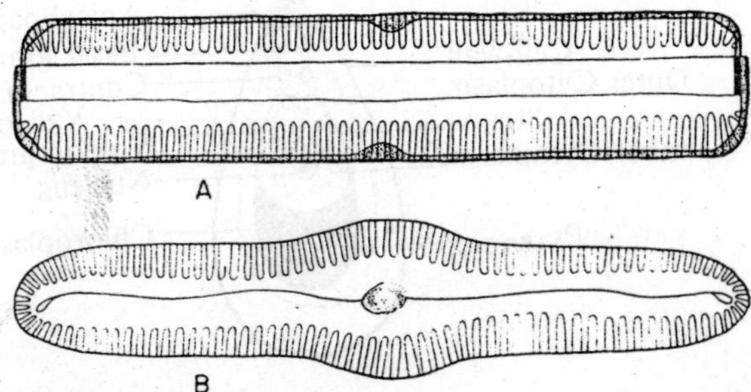

Fig. 2.23. A side view (A) and a top view (B) of a typical diatom, highly magnified.

red, brown or blue-green may develop in some groups apart from the presence of chlorophyll. Algal classification as such is based on type of pigmentation and not on structural diversity and study of algae is called *phycology* or *algology*.

Algal Classification

Divided into 5 classes based on pigmentation.

Cyanophyta

They are most primitive among other groups of algae, with no distinct nuclear structure and are often considered as cyanobacteria that have many similarities with bacteria. The pigment is blue-green *phycocyanin*, they form major part of phytoplankton and are free floating eg., *Oscillatiria*, *Anacystis*, *Microcystis*. Some are terrestrial growing on moist soils, eg., *Anabaena* and *Nostoc*, which help fixation of atmospheric nitrogen in soil. In *Oscillatoria* filament, cells are cylindrically attached, while in *Nostoc* and *Anabaena*, the filaments are beaded with round or elliptical cells. In many eutrophic lakes, *cyanophycean bloom* is observed.

Chlorophyta or green algae

This is the largest group of algae and common in fresh water and are diverse shapes and sizes from unicellular bacteria like motile *Chlamydmonas* to small non-motile round cells forming colonies as in *Chlorella*, to ball like colonies of *Chalmydomanas* like cells as in

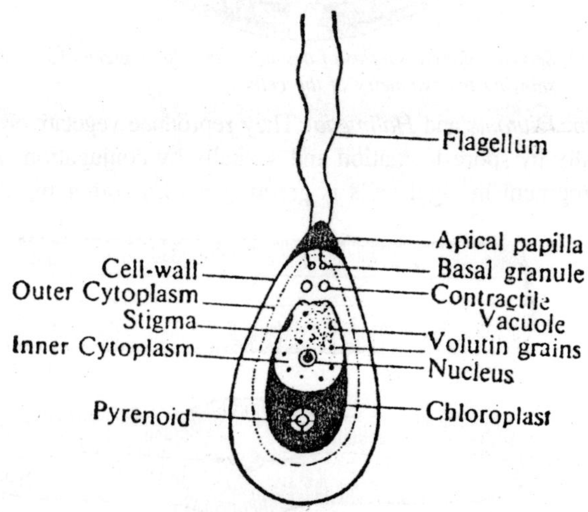

Fig. 2.24. Chlamydomonas.

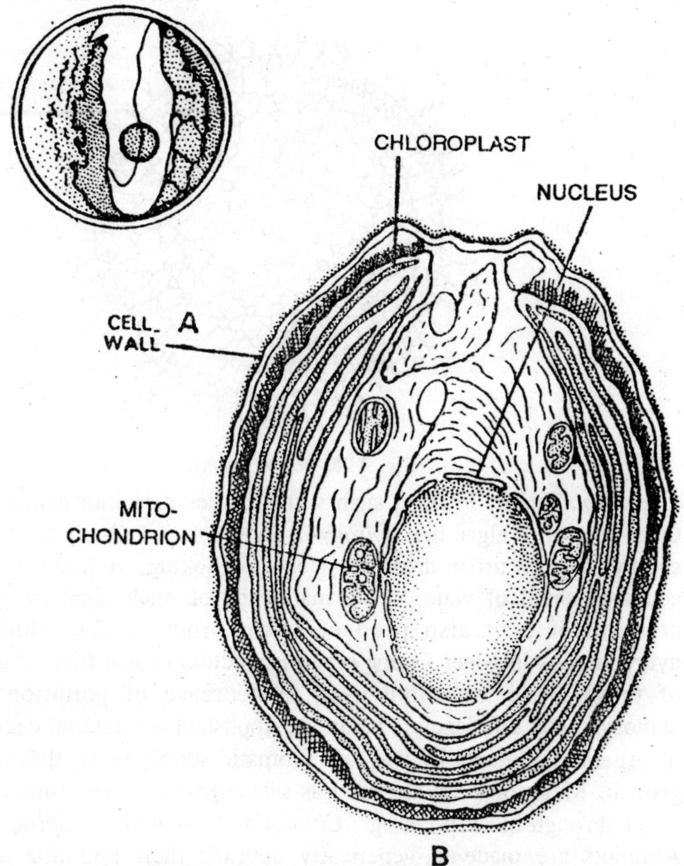

Fig. 2.25. Chlorella. A—Under ordinary microscope: B—Under electron microscope.

Volvox, to long filamentous types of *Spirogyra*, *Ulothrix*, *Oedogonium*, to net-like structures in *Hydrodictyon*.

They are dark green and reproduce either by spore formation or sexually by conjugation. *Spirogyra* is the common representative, where conjugation is ladder-like i.e., 2 filaments come together and through small peg-like lateral protuberences, contents of one cell pass into the other and fuse.

Most chlorophycean algae rich sources of protein are important and their mass cellulose is done in many countries to make food products as in *Chlorella* and *Spirulina*. In motionless water, which is rich in organic matter, large-scale growth of various members of this group can cause waterbloom and overgrowth and later their decay may lead

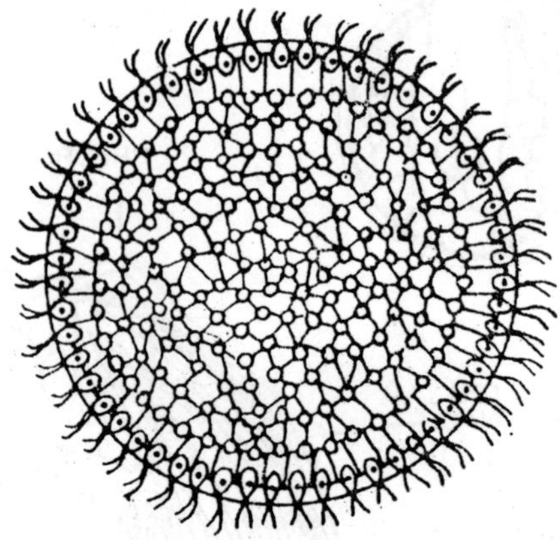

Fig. 2.26. Volvox colony.

to *eutrophication*. Finally, such water bodies turn into marshy lands. Over growth of algae like *Spirogyra*, *Zygnema* and *Oscillatiria* in water supply pipes in urban areas may lead to choking, and their death can add foul smell of water. Chlorination check such algal overgrowth, however it may also produce toxic products like chlorinated hydrocarbons in water Ozone (O_3) disinfection is also tried. And many of these algae are often used in decrease of pollution. Their photosynthetic efficiency is being accomplished in bacterial degradation of organic pollutants by way of automatic supply of O_2 through algal growth. Many of them are used as scavengers of toxin minerals from water through biosorption eg., *Chlorella*, *Scenedesmus*, *Spirulina*, etc. Attempts are made to genetically upgrade their pollution decrease potential, besides number of chlorophycean genera are tried for biological hydrogen generation as an alternate source of fuel.

Some of common pollutant tolerant algal genera :

1. Chlorophyta	*Chlamydomonas*, *Euglena*, *Chlorella*, *Scenedesmus*, *Spiroggra*, *Ulothrix*, *Pandorina*, *Hydrodictyon*, *Cosmarium* and *Stigeoclonium*.
2. Cyanophyta	*Anabaena*, *Oscillatoria*, *Spirulina*, *Anacystis*
3. Bacillariphyta (Diatoms)	*Pinnularia*, *Navicula*, *Hantzchia*, *Nitzschia*
4. Phaeophyta	*Fucus*, *Ulva*, *Codium*, *Sargassum*

Many algal species and genera are taken as good *indicators* of organic pollution. Palmer (1969) for the first time emperically developed

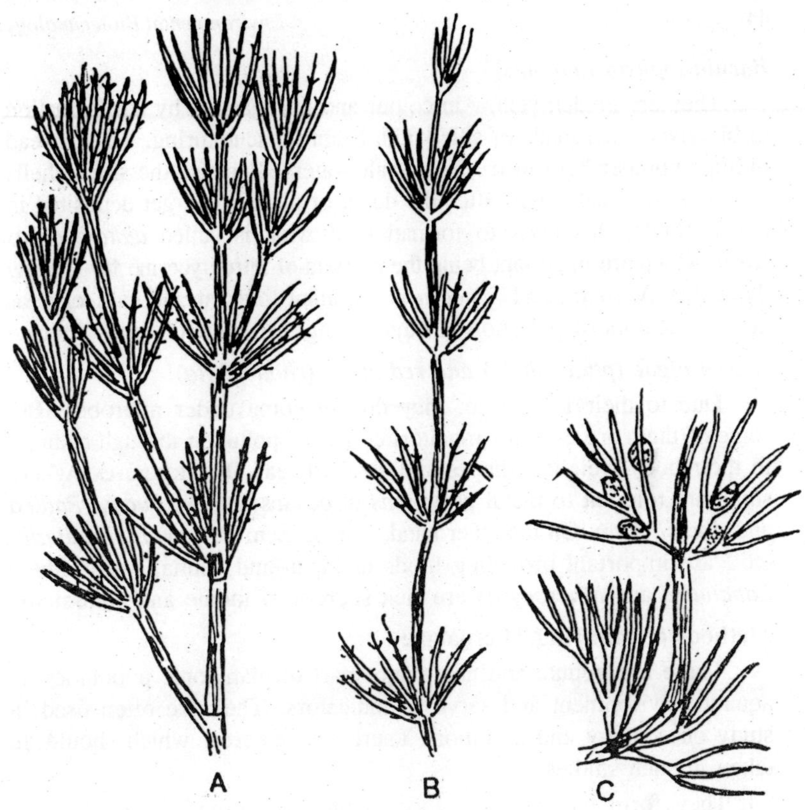

Fig. 2.27. Chara.

pollution indices of various algal genera, based on the proposal of counting the number of individuals of a genus present in one ml. of polluted water. A score of 20 or more, obtained by adding the values of each genus present, indicated organic pollution. On this basis, the *pollution indices* of the genera have been worked out, which are as follows:

	Genera	*PI*
1.	Euglena	5
2.	Chlamydomonas	4
3.	Oscillatoria	4
4.	Navicula	3
5.	Scenedesmus	4
6.	Pandorina	1
7.	Stigeoclonium	2
8.	Chlorella	3

Bacillariophyta (diatoms)

They are golden yellow in colour and are features by the formation of bi-valved shell made of silica with beautiful sculpturing. They spread in huge number both in fresh and salt waters. Because the silica shells withstand to breakdown, after the death of algae, they get deposited in marine beds, this leads to formation of what is called *diatomaceous earth*, which are important being the carriers of nitroglycerine for making dynamite. *Navicula* and *Synedra* is pollutant tolerant and can serve as good indicators of pollution through changes in their diversity index.

Brown algae (phaeophyta) and red algae (rhodophyta)

Due to their large size, they do not come under microbes. But some of them are used for monitoring marine pollution through changes in their body weights, shapes and sizes of leaf-like structures, where some are tolerant to metal pollutions in oceans namely *Fucus*, *Padina* and *Ascophyllum*. On the other hand, few of them *Porphyra*, *Gracilaria*, etc., are important providing foods in Japan and China, while others *Laminaria* and *Macrocystis* are rich sources of iodine and potassium.

Methods of Studying Microalgae

These are minute and take major part of planktonic population in aquatic environment and serve as indicators. They are often used in study on ecology and pollution. There are features, which should be taken in such studies.

1. They form the base of 'good chain' and provide food to other organisms.
2. They can be easily cultured with well-established culture techniques, and
3. They are countless (Ca. $10^2 - 10^6$ cells/ml) and thus, with a very small volume of water such studies can be done. The counting of microalgae in a sample population is done by any one of the following three methods.

Drop method

Refers to number of organisms in drop of water

$$n = \frac{\text{Area of cover glass}}{\text{Area of one microscopic field}} = \text{Average count per field.}$$

$$\text{If } n = 10, \text{ then } x = \frac{22\text{mm}}{0.01\text{mm}} \times (n = 10).$$

where x represents total number of organisms in a drop of water, while 22 mm is area of a cover glass and 0.01mm is the area of one microscopic field.

If there are 5 drops in a ml, then organisms per ml = X × 5 =?

Haemocytometer method

This is used in blood cell counting, and is a counting chamber on a glass slide and has centrally located 'H' shaped groove with cubical chambers and grooves are 1 mm × 1 mm × 0.1 mm in area, while the central chamber is divided into 25 sub-divisions.

Sedgewick rafter cell method

It is simple and reliable which consists of a glass slide of 50 mm × 20 mm × 1 mm deep and volume of the S.R.Cell is 1 cm^2 (1 ml).

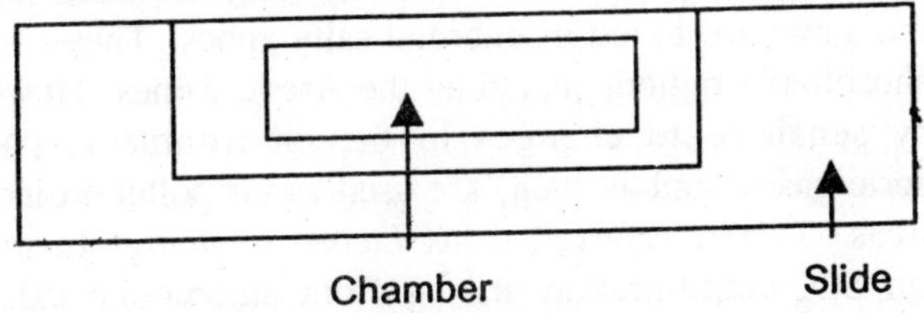

Fig. 2.28. Sedgewick rafter cell.

Planktons are allowed to settle for sometime and then counted under the microscope.

$$\text{No. of Plankton / ml} = \frac{\text{No. of organisms counted}}{\text{No. of replications taken}}$$

LICHENS

A lichen represents a permanent and unique association of fungal and can algal component in its body to appear as a single individual. The 2 parts are mycobiant and phycobiant, though lichen appears as a single macroscopic organism, yet individually, each component is microscopic, i.e., made up of 2 different microbes.

The word 'lichen' is of Greek and means as a scaly superficial growth on tree barks, and usually grow as whitish waste on base rocks, tree barks and leaves of other plants where the colour may be black and also red.

The main body of lichen is made of fungal hyphae of ascomycetous group which enclose free unicellular algae of green or blue-green. Although is often regarded as symbol of *symbiosis*, yet in this the algal part provides food, while the fungal one givens only shelter and retains moisture for proper growth, this enables the association to survive in such environments where individual component could not have thrived along and they grow extremely slow.

On structural formations, they are grouped as (a) *foliose* having a leaf-like thallus (b) *crutose* which grow as crusts on rocks, tree branches (c) *fruticose* that are larger, pendulous and hang losely from the supporting substrates.

Common members are *Usnea, Cladonia, Rocella, Parmelia, Labania* and *Graphis*. Many are important and some provide a typical starch like carbohydrate called *Lichenin*. Usnic acid is an alkaloid which has antibacterial property. The red dye *orcein* is got from lichen, while acid/base indicator *Litmus* is got for testing pH of water. On the other hand, a French perfume is extracted from species of *Lobaria* and *Evernia*. Lichens are real pioneers in the world of ecological succession and they established on bare rocks. Though they are often found in dry areas, and some are found in interdial salty zones. They can thrive in many inhospitable regions including the Arctic Zones. However, they are very sensitive to changes in the environment, particularly atmospheric gases, and as such, are used as air pollution indicators of urban areas. This is referred to as *Lichen mapping*. In urban areas with high SO_2 concentration arising from automobile exhaust, they either disappear quickly or their frequency of distribution gets reduced Hawsksworth (1971). Now it has become a practice to have lichen mapping of cities like London, Mumbai and Kolkata as bioindicator of air pollution. Lichens also have ability to scavenge minerals through the excretion of a chelating called *lichen acid* which can chelate and remove metals from the environment.

PROTOZOA

This is primitive group of animals and belongs to protists. *Proto* means primitive and *zoon* is animal and the scientific discipline is known as *Protozoology*. They are unicellular, microscopic, eukaryotic, however they are much larger than bacteria and the size varies from 5 μm to 250 μm in diameter which occur in colonies or single.

Features

1. They may be aquatic, mostly found in moist habitats, including sewage, sludge and excrements.
2. There are about 3000 species, most of which are free-living, while some are parasitic to animals and men, causing millions of deaths worldwide. The pathogenic forms have been listed in the beginning of this chapter.
3. The population controlling factor : Light, temperature pH and nutrients. Optimum temperature is 16^0–25^0C and the maximum temperature between 35^0–45^0C.

4. Nutrition is *holozoic*, which means the absorption of soluble food taken place throughout the body surface. Food may also be engulfed and digested in *food vacuoles*, while excretion taken place through *contractile vacuoles*.
5. Organs of movement are pseudopodia cilia and flagella.
6. They may be spherical, radical or bilateral.
7. Reproduction is by binary division, while some reproduce sexually by conjugation, can withstand unfavourable environmental conditions by thick-walled cyst formation.

Many of them could be pathogenic to men and animals either as ecto- or endo-parasites, they grow abundantly in waste water treatment facilities. They may help cleaning the water by injecting harmful bacteria and used as indicator organisms for assessing toxicity levels of water, eg : *Tetrahymena* and *Paramoecium*. High amount of protozoa indicates healthy condition of a sludge.

Classification

Divided into 4 on basis of their structures.

Rhizopoda or Sarcodina

They are featured by the presence of *pseudopodia* which protrude out of slimy body during gliding on a surface and injestion of food. There is no spore formation and reproduction takes place by fission. The whole body may be converted into a hard wall during unfavourable environmental conditions including drugs. *Amoeba* is a typical member of group with simple, soft cell having large nucleus, food and contractile vacuoles. The common amoebic dysentery is caused by *Entamoeba histolitica* which also cause liver abscess. Another is *Naegleria fowleri*, which is associated with pollution and can cause meningo-encephalitis leading to death.

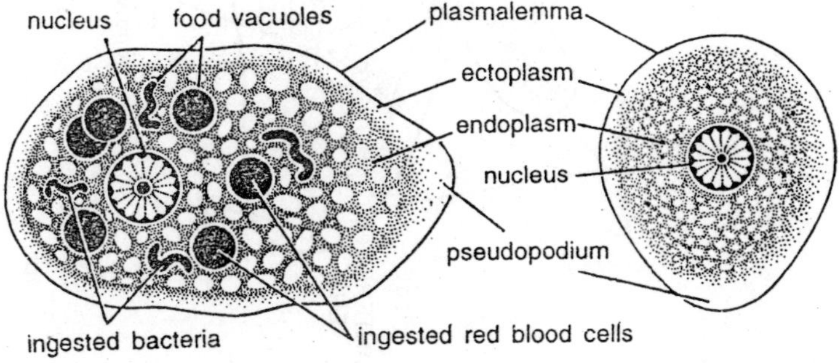

Fig. 2.29. Entamoeba histolytica.

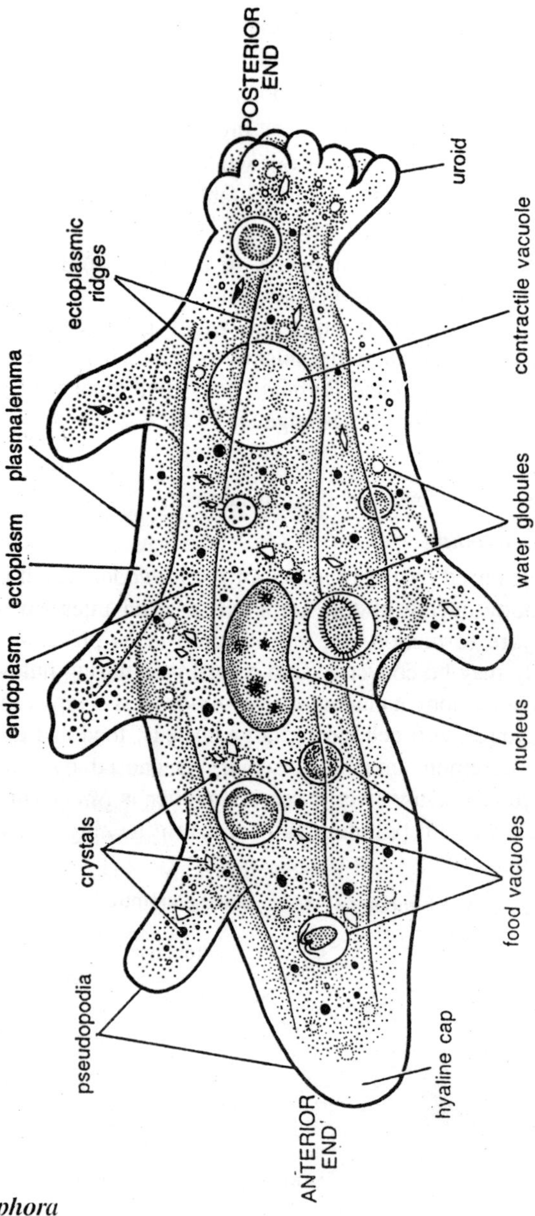

Fig. 2.30. Amoeba proteus.

Mastigophora

They are flagellated protozoa, which have whip-like structures for moving and no spores are formed and their division taken place

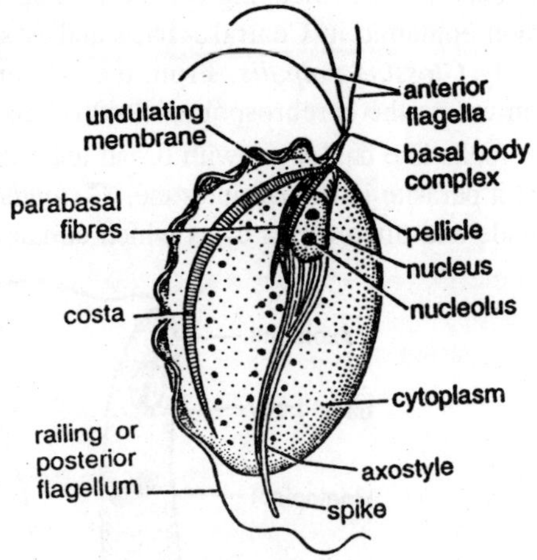

Fig. 2.31. Trichomonas tenax.

longitudinally. Sexual reproduction in some groups only while encystment is common to all. They are frequently insect-borne pathogenic parasites. Some members of the group are of pathological importance are discussed here.

(i) *Trichemonas*. has egg-shaped body with four flagella, one which is long, it is a parasite in the cavities of vertebrate animals. *T. vaginalis* infects human vagina, while *T. hominis* cause mild diarrahoea.

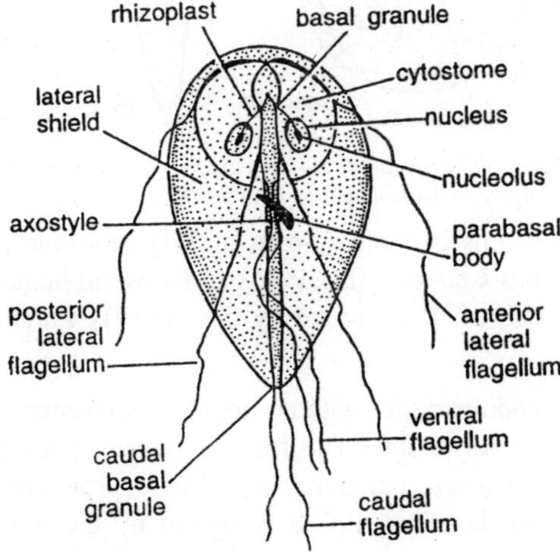

Fig. 2.32. Giardia lamblia.

(ii) *Trypanosoma*. Slender wave shaped flattened-body with pointed ends *T. gambiense* causes 'Sleeping Sickness' disease in man which is a common epidemic in Central Africa and is spread by the blood sucking fly *Glossina palpalis*. From the salivary gland of fly, the pathogen enters the cerebrosphinal fluid of man.

(iii) *Giardia*. looks like half-spear with broad anterior end and 4 flagella which is a parasite in human intestine. *G. lamblia* causes diarrhoea world wide and often forms cysts which contaminate water bodies.

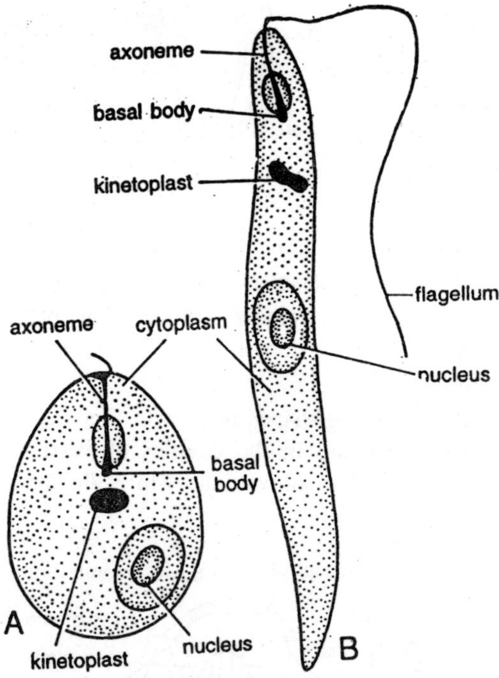

Fig. 2.33. Two forms of Leishmania. A—Amastigote or leishmanial form in man. B—Promastigote or leptomonad form in sandfly.

(iv) *Leishmania*. this is another form of *Trypanoosoma*, which causes the dangerous *kala-azar* (black fever) disease in human beings when skin gets darker. It is spread by the sand fly *Phlebotomus*.

Sporozoa

These are endo-parasites with no organ for movement and reproduce largely by spore formation. The most important pathogenic genus is *Plasmodium* or malaria parasite, its sexual reproduction taken place inside the gut of the mosquito. It is spread by the female *Anopheles*

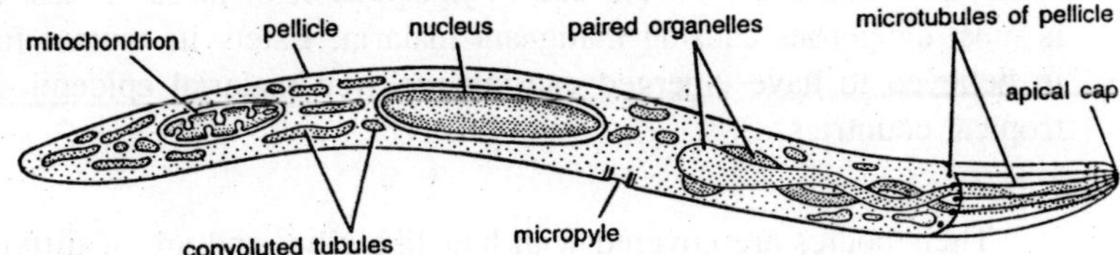

Fig. 2.34. Plasmodium. Structure of a sporozoite as revealed by electron microscope.

mosquito. Part of its life cycle is completed within the body of mosquito and part in human blood.

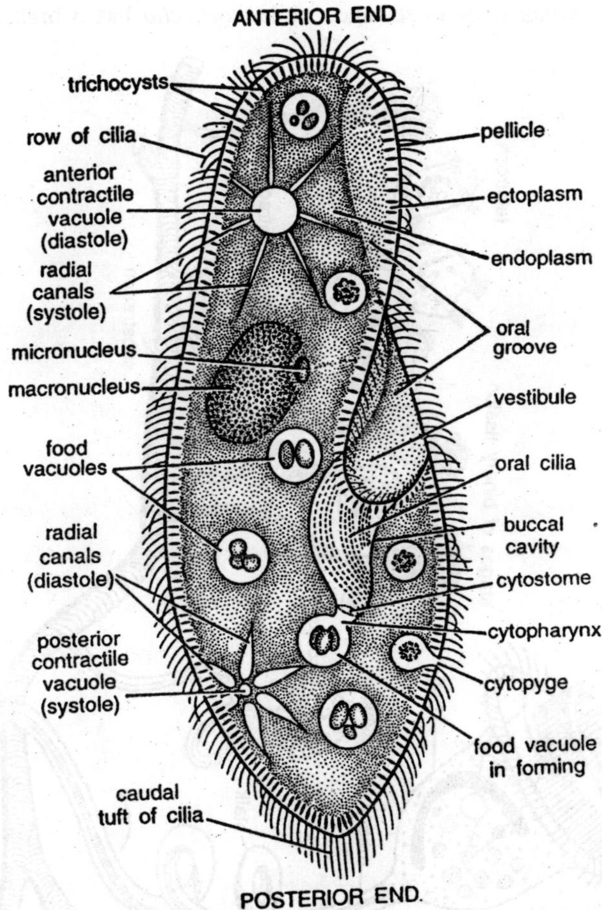

Fig. 2.35. Paramecium caudatum.

Plasmodium has oval shaped body and has 4 species namely *P. vivax*, *P. malariae*, *P. ovale* and *P. falciparum*, of these the last one is most dangerous causing malignant malaria. Lately its mutant form in believed to have emerged causing renewed malarial epidemics in tropical countries.

Ciliata

Their bodies are covered with hair-like cilia, and are of different shapes and sizes viz., elliptical, pear-shaped, bell-like, etc.

Balantidium causes mild diarrhoea and ulcer in colon *B. coli* is an intestinal parasite of hog and infection is transmitted to pork-eating persons. The representative of this group is *Paramoecium* protozoan. *P. caudatum* lives in sewage, while *Vorticella* has a branched, bell-

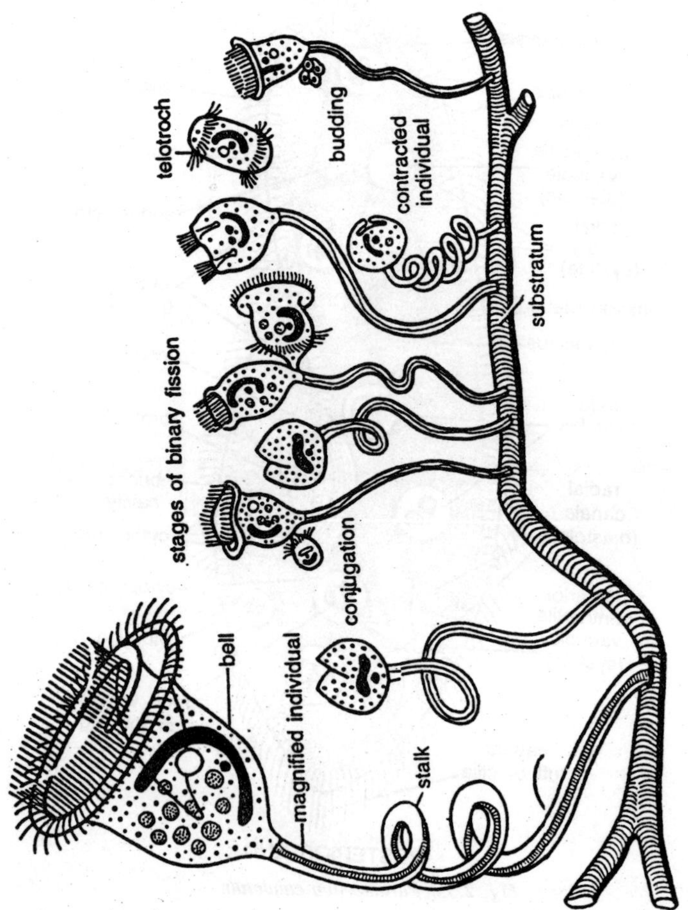

Fig. 2.36. Vorticella. A group of individuals in different states.

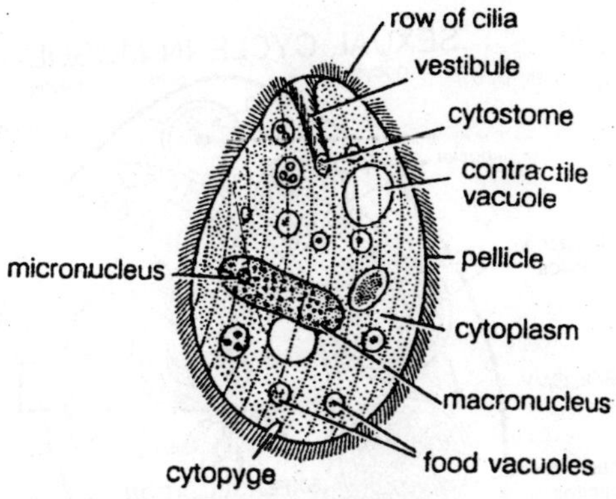

Fig. 2.37. Balantidium.

like structure and is found in stagnant water. Another member is *Tetrahymena* with a ciliated pear-shaped body. *T. pyriformis* is often used as indicator organism in water pollution studies.

DAPHNIA

Besides the protozoan group, another microscopic organism belonging to insects is *Daphnia monga* which is very responsive to

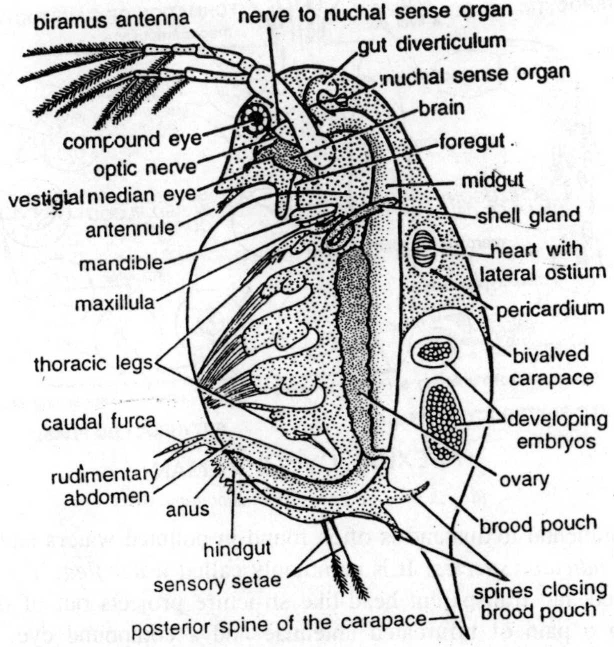

Fig. 2.38. Daphnia. Female in lateral view.

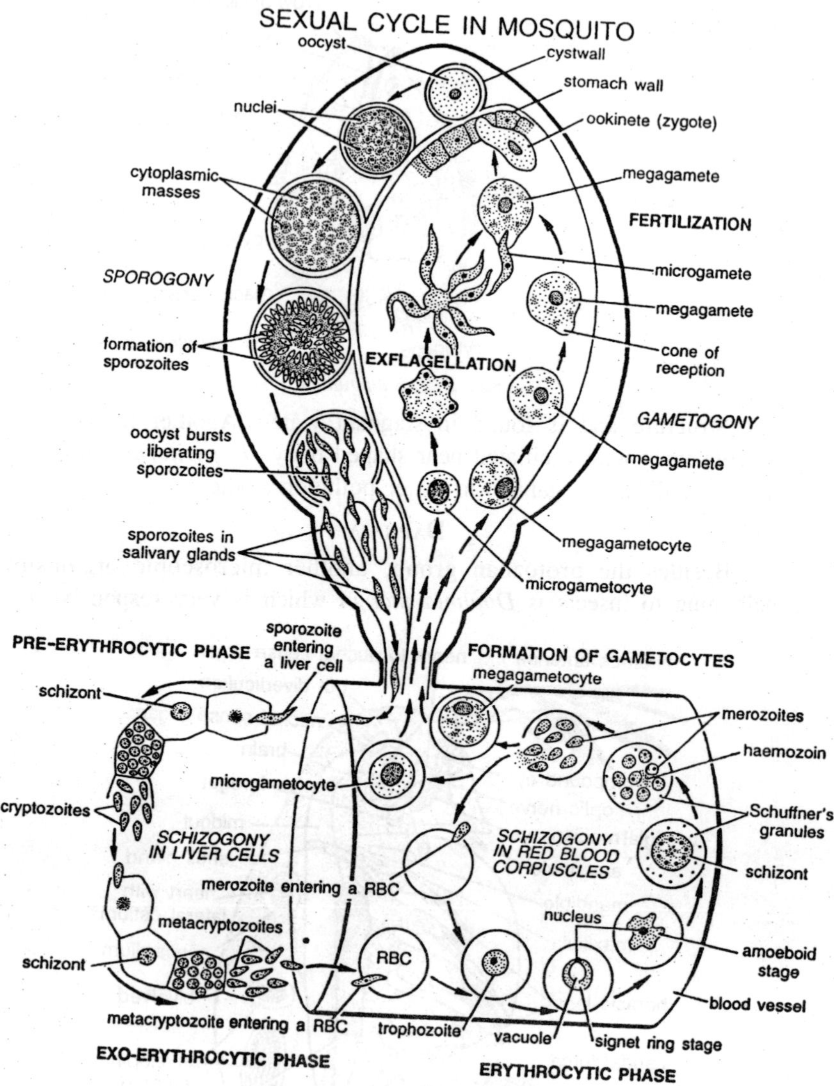

Fig. 2.39. Life cycle of Plasmodium vivax.

environmental toxins and is often found in polluted waters and is used as an *indicator species*. It is commonly called *water flea*. The body is bivalved and transparent head-like structure projects out of the body having a pain of bifurcated antennae and a compound eye. Female water flea has a 'broad-pouch' filled with eggs.

3

WATER POLLUTION

During the last decades, parallel with rapidly developing technology, increasing populations and urbanization we have been witnessing alarming phenomena all over the world. In almost every country air, water and soil pollution, the decrease in arable land, the danger of radiation, the accumulation of solid wastes, the depletion of energy carriers and of mineral resources, the death of parts of the plant and animal kingdoms have been becoming dominant problems. The problems arising from increasing urbanization include the obsolescence of the infrastructure, the pollution of the city atmosphere, the lack and bad quality of the drinking water, overcrowding and traffic difficulties, the decrease in the areas of greenery, parks and generally the so-called damage done by civilization. Different contaminants (pesticides, oil, sewage, faeces, detergents) and excessive use of fertilizers affect both the living organisms of the soil and the vegetation of higher plants.

Approximately 900,000 t pesticides are applied yearly. Upto now some 80,000 t DOT has accumulated. DOT can be detected in the entire living world and can also be found in significant quantities in the human body. The growing load of environmental pollution gives rise to an increase in the number of so-called civilizational damage, that menace the whole of mankind. Because of the manifold and increased overstressing of the human organism, gastric ulcers, cardiac infractions, chronic illnesses of the respiratory system, neurosis, etc., occur more frequently. Often different chemicals get into the human organism.

The occasional appearance of certain unidentifiable illnesses, like the Minamata disease or the Itai-Itai-disease can also be traced back to the toxicity of chemicals. Parallel to pollution of the environment

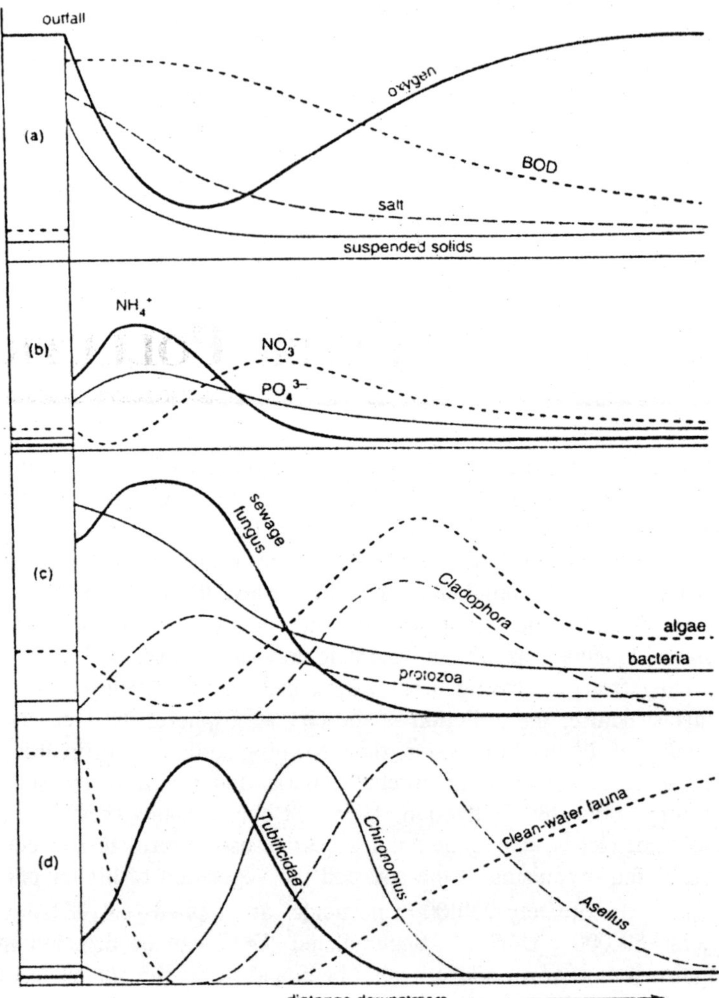

Fig. 3.1. Schematic representation of the changes in water quality and the populations of organisms in a river below a discharge of an organic effluent (a) represents physical changes, (b) represents chemical changes, (c) represents changes in micro-organisms and (d) represents changes in macro-invertebrates.

and of the living world (air, water, soil) have been subject to environmental stresses and a demographic explosion may have to be reckoned with, around the years 2000. According to surveys of the UNO the world population will grow from 3 to 4-6 billion between 1960 and 2000. All these factors point to the danger of the upsetting of the natural environment. Others, on the contrary, express the view that with the appropriate means for the control of population growth in the hands of humanity the demographic explosion could be evaded.

Water Resources of the World

According to our present knowledge the total water resources of the world amount to 26.6 trillion t. Approx. 94.7 percent of this huge volume of water can occur in the lithosphere (rocky belt), its major part being bound to minerals which constitute the rock bed. This is known as bound water (other names: water of crystallization, water of constitution). These waters have been forming a part of the structure of minerals and get released only at high temperature. The remaining 5.3 percent can practically be found in the hydrosphere (water belt). The oceans and seas form the greatest coherent volume of water to be found on the surface of the Earth. They have been covering 70.84 percent of the 510 million km^2 of surface area of our planet. Their depth exceeds, in more than one location, the height of the highest peaks of the continents.

On the surface of the continents water appears in a more scattered form, covering 2.5 million km^2 of its territory. From this, the area of fresh water amounts to 2 million km^2. The volume of fresh water has been small in comparison with that of seas and oceans (Table 3.1). It amounts to barely 0.4 percent of the surface area of the Earth and to approx. 1 percent of the area of the continents.

Table 3.1. Water Resources of the Hydrosphere

Location and state of stored water	*Amount 10^{12} t*	*Percent*
Seas	1,380,000	98.900
Potar and mountain ice and snow	16,700	1.077
Fresh water	25	0.002
Water vapour in the atmosphere	13	0.001
Underground water	250	0.020
Total	1,396,988	100.000

In the Earth's interior, water is found in various forms of underground water. Capillaries of granular soils are filled with coherent soil water. Soil binds water by means of capillary action (soil moisture), though the existence of subterranean streams should also be mentioned. The quantity of biologically bound water, in comparison with the above forms, has been somewhat insignificant. The biologically bound water implies water that forms part of the structures of plants and of the bodies of animals. The so-called transpiration water, transpired by plants in the course of their metabolism should also be considered. In the form of invisible vapour and the condensed water of clouds, finely distributed water has been also present in the atmosphere.

Degradation of Water Resources

Increased water consumption caused by urbanization and industrialization results in the lowering of the underground water table. For example, in Italy, around Milan, the underground water table has sunk by 20 m, over a period of 20 years. In 1945, in Bologna, underground water could be found at a depth of 12 m; today it is to be found at 45 m. Industrial effluents and domestic sewage, detergents, pesticides and oil have been all polluting the rivers, lakes and seas.

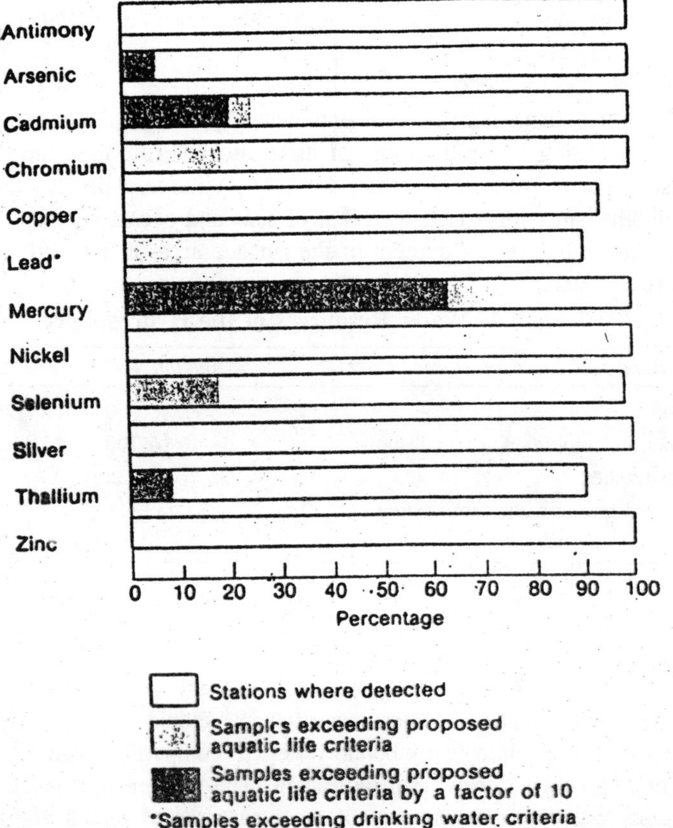

Fig. 3.2. Inorganic toxic pollutants detected in a survey of river water quality.

The death of living organisms in Lake Erie has been already known. There have been other alarming reports on dying seas (the Baltic), on the damage caused by oil-spills on marine-living organisms, on the Rhine having become a sewer, etc.

The pollution of marine waters has been regarded as a matter for special concern as the seals phytoplankton and particularly algae have been the world's largest oxygen-producing resources. Indications of water pollution have been multiplying in Hungary, too. To mention only a few: the river Sajo has becomes a lifeless water way, fish kills occur in Lake Balaton and periodically, polluting substances of smaller or greater quantity emerge in the rivers, bathing is forbidden along the Danube reaches of greater Budapest, etc. Because of the spend of newly established industrial plants, towns, resort-areas, water and gas pipelines, highways and roads, the area of arable land and that of natural plant cover, forests, meadows and marshes has decreased. Because of erosion and salinisation, approx. 500 million ha of arable land has become infertile in the world. One quarter of the arable land in China is eroded and 2.5 billion t of arable land is lost every year. The river Rhine carries 4.6 t of slit into the sea. According to a report from the UNO, more than 10 million ha are lost from agricultural cultivation because of secondary salinisation. In Hungary, 150 million t of fertile soil has been carried away by rainwater.

Classification of Natural Waters

Natural waters which have been comprising all the waters of the continents have been divided into two groups: seas and inland waters. Besides their vast extent and volume sea water can be characterized by a mean 3.5 percent dissolved solid content of which 2.73 percent is salt (sodium chloride). The volume of inland waters has been less, their salt content being under 3 percent. The majority of them has been so-called fresh water, with a dissolved solid content under 0.05 percent. Waters can be classified according to several principles like organic matter content, temperature conditions, turbulence, vegetation, etc.

Particularly in the case of lakes, numerous classifications are used. From among them Thienemann's (1926) classification can be considered as the most comprehensive:

(i) Ground waters
(ii) Springs
(iii) Running waters (from brooks to rivers)
(iv) Standing water: lakes, lagoons, pond or swamps, temporary waters

(v) Waters of unusual temperature and chemical composition: thermal waters, snow and ice, sewage, brackish waters (with a salinity of 0.05-3 percent), highly-saline and natron (chemically rich) lakes and other waters having peculiar characteristics.

On the basis of the use of water domestic, industrial and agricultural sewage can be distinguished. These have been very different in their composition and even within a certain group great differences can be observed concerning their content. As to the biological impact of substances to be found in sewages, they could be divided into two main groups: nutrients and biologically active substances, i.e., poisons. Organic matter utilizable by heterotrophic organisms (bacteria, worms,

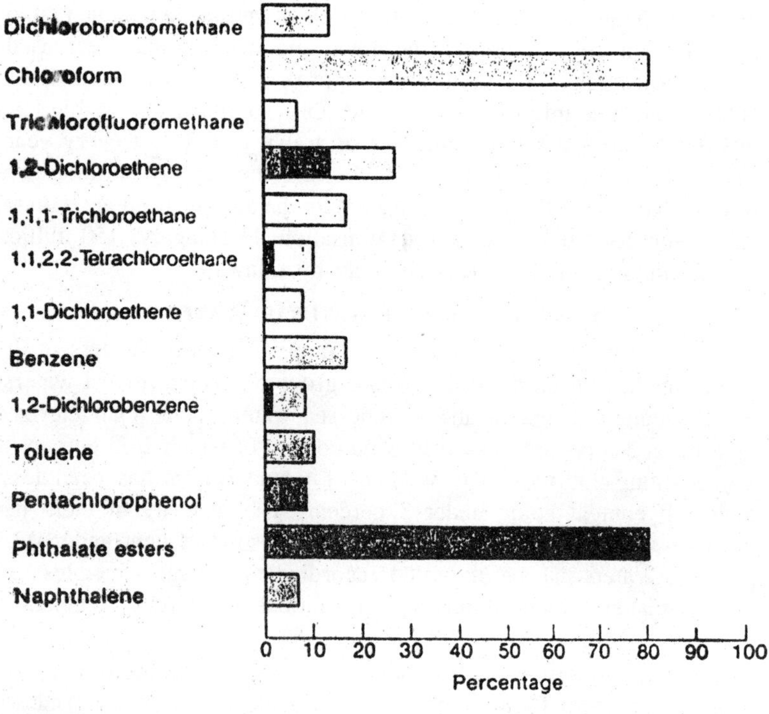

Fig. 3.3. Organic toxic pollutants detected in a survey of river water quality.

arthropods, etc.) form the first group of nutrients, whereas the inorganic nutrients utilized by autotrophic organisms (algae) belong to the other one. In this context it should be mentioned that there have been two important water quality indicators:

(i) The accumulation of decaying organic matter available for heterotrophic organisms result nothing else but saprobity.

(ii) The amount of inorganic (mineral) substances available for algae represents the degree of trophicity.

Biologically active substances (poisons) have been forming two main groups. Group 1 is having inorganic matter (microelements: fluorine, zinc manganese, etc.: active residues; NO_3; different gases, etc.), whereas organic matter (detergents, plastics, pesticides, phenols and different kinds of tar) belong to Group 2. The toxic effect of waters, which can hinder or even prevent, aquatic life, is expressed by toxicity as an index of water quality. Sewage modifies the water quality of surface waters. It is established that downstream of the source of pollution, after a while, polluting matters gradually disappear and water regains its original state. This process is termed as self-purification. It has been the result of physical, chemical and biological processes in which living organisms have been playing the most important role. From among the physical factors, water movement, viscosity, temperature and light condition can be underlined.

Hydrolysis and oxidation of polluting matters have been the most important chemical processes, interrelated with the metabolism of life communities, which take place in two phases. In the anaerobic phase (decomposition and actual decay) bacteria are able to decompose organic matter in the absence of oxygen. In the aerobic process, in the presence of oxygen besides bacteria also plant and animal organisms take part in decomposition. It has been however, impossible to determine the exact boundaries between the two processes.

Table 3.2. Forecast Water Demand and Wastewater Discharges (Million m^3/year)

	1965	*1970*	*1975*	*1980*	*2000*
Total water demand	3.700	4.800	6,400	8.700	10,500
Fresh water demand	1.600	2,100	2.800	3,900	5,800
Sewage discharges	390	500	665	900	1,200
Fully purified sewage	70	120	330	630	930

Aquatic living organisms have been studied by numerous scientists. It is well established that the occurrence of certain organisms has been an indication of a particular degree of pollution. This can be explained by the fact that aquatic organisms have adapted themselves to the varying degree of pollution and to the differing chemical conditions derived from it.

Among the organisms of the plant and animal kingdom, the single celled animals (Protozoa) have been of special importance, though there have been also other organisms which render it possible to assess the degree of water saprobity. In his classical work Leibmann (1962) distinguishes 4 degrees by which he characterizes the state of pollution by domestic sewage (Table 3.3).

Table 3.3. Characterization or the Degree or Water Pollution

Degree of pollution	*Saprobity*	*Colour of water*
Very severely polluted	Polysaprobic	Red
Strongly polluted	α-mesosaprobic..	Yellow
Moderately polluted	β-mesosaprobic	Green
Slightly polluted	Oligosaprobie	Blue

The polysaprobic type of water has been highly polluted with a great amount of protein, high molecular weight N-containing compounds, polypeptides; grease and their decomposition products. It is having an offensive odour. As oxygen is generally absent, the process of self-purification starts under anaerobic conditions. At the beginning of decomposition there has been a great number of bacteria present. The α-mesosaprobic type water has been characterized both by the increase of dissolved oxygen and by the presence or amino acids derived from the decomposition of proteins. Besides the still great number of bacteria, the presence of algae has been also significant.

The oxygen demands has been still great, though the water does not smell so badly as polysaprobic sewages do. Waters of the β-mesosaprobic type get characterized by a significant oxygen content for which there has been only little oxygen demand. While the number of bacteria decreases, it has been rich in other species and individuals of microscopic organisms. This life community has been very sensitive to the impacts of sewage. The oligosaprobic water has been possessing a great amount of dissolved oxygen, while the number of bacteria has been still insignificant. It is rich in species but poor in individuals.

This life community reacts to changes in the composition of water very sensitively. Its members have been not capable of enduring poisonous substances even in very small concentrations. It has been not only the quality and quantity of living organisms that renders it possible to get information on the degree of water saprobity but also the chemical characteristics of the water (quantity of organic matter, biochemical oxygen demand, processes in connection with the oxygen cycle, etc.) Due to the use of the agricultural use of chemical, pesticides, which belong to the class of biologically active substances, have become one of the most widespread poisons.

According to the type of organism to be killed, nearly 200 poison types have been known (acaricide-mite, larvicide (mosquito), nematicide (roundworm) killer, etc. Concerning their active ingredient they have been most varied (zinc and arsenic compounds, chlorinated derivatives, phenoxy derivatives, DDT and its derivatives, etc.). The number of the most frequently applied poisons exceeds 60. Numerous authors described tragic cases in foreign countries where lakes and rivers suffered from pesticide kills. These substances could be able to penetrate into the living space in three ways:

(a) by direct treatment of the water surface;
(b) by wind, from the treated arabie land;
(c) from treated land, by way of water infiltration.

Data in the international literature appear to indicate a tendency towards the application of pesticides with a short time of decomposition. In this way the amount of damage can get reduced. Certain pesticides persist in the soil for a very long time without losing their action. Thus, e.g., DDT, methoxychlorine and penta-chlorophenol persist for 15 years, aldrin, dieldrin, BHC (benzene hexachloride), heptachlor for 9-12 years, toxaphene (camphecor) for 5 years.

In spite of certain objections it seems to be proved that the remnants of persistent herbicides have been carried into waters from land either in the form of wind-blown dust and washed-off deposits or without it, in diluted or suspended form. The large-scale fish kill in lake Balaton, in 1965, was also caused by DDT and γ-BHC (Lindane). While these components were hardly detectable in water the organisms serving as food for fishes contained, on the contrary, large amounts of them. Therefore it could be concluded that the accumulation of poisons had occurred along the food chain (algae→ invertebrates→non-predatory fish→predatory fish). Due to the widespread use of pesticides it has been a very complex and difficult task to protect against them. Sufficient protection can be by the sensible choice and application of pesticides though the quantity of chemicals applied should also reduced.

According to Szabo (1973) in 1971, in the catchment of Lake Balaton alone, in large-scale agricultural production 994.3 t of pesticides has been used up. On the basis of preliminary investigations, it can be concluded that the Lindane pollution of water courses has been widespread in this region. A residue of 2, 4-0 could also be detected. Synthetic detergents are having large molecular compounds. The exert their action either directly or indirectly Frequently used detergents are poisonous for a very large number of aquatic organisms. The production and use of synthetic detergents is rapidly increasing, therefore, their danger must be taken into serious consideration. From among other pollutants, mineral oil should be mentioned. The pollution of rivers occurs most frequently.

Water Pollution

The great solvent power of water has been making the creation of absolutely pure water a theoretical rather than a practical goat Even the highest quality distilled water is having dissolved gases and to a slight degree solids. The problem, therefore, has been one of determining what quality of water has been required to meet a given purpose and then finding practical means of achieving that quality. The problem gets further complicated because every use to which water has been put—washing, irrigation, flushing away wales, cooling, making paper, etc., has been adding something to the water.

In fact, for centuries rivers and lakes have been used for dumping grounds for human sewage and industrial wastes of every conceivable kind, many of them have been highly toxic. Added to this have been the materials leached and transported from land by water percolating through the soil and running off its surface to aquatic ecosystems. The term 'water pollution' is referred to the addition to water of an excess of material (or heat) that is harmful to humans, animals, of desirable aquatic life, of otherwise causes significant departures from the normal activities of various living communities in or near bodies of water. The National Water Commission stated (1973) that "water gets polluted if its has been not of sufficiently high quality to be suitable for the highest uses people wish to make of it at present or in the future. In reality, the term water pollution refers to any type of aquatic contamination between two extremes:

1. a highly enriched, over productive biotic community, such as a river or lake with nutrients from sewage or fertilizer (cultural eutrophication), or
2. a body of water poisoned by toxic chemicals which eleminate living organism or even exclude all forms of life.

The expression "water pollution" seems to be clear to all. Nevertheless, it is worth determining its real meaning as this has changed in the course of time. Felfoldy's (1972) precise definition is the following: "Water pollution is every impact which changes the quality of our surface and subsoil waters to such a degree that its suitability either for human consumption or for the support of man's natural life processes will decrease or cease." Most of the wastes of civilization are drained into streams and rivers and ultimately into lakes or oceans. In many cases the waste is dumped into the same bodies of water from which drinking water is withdrawn.

The effect of sewage, industrial waste and agricultural drainage on plant and animal life in closed bodies of water is sometimes catastrophic. The accumulation of excessive plant nutrients, called *eutrophication*, has accursed in Lake Erie, Lake Washington at Seattle, the lakes in the Madison, Wisconsin, area, and Lake Zurich in Switzerland. Coastal waters are also changing due to organic material from sewage outfalls. Marine algae are disappearing from sections of the California coastline. Kelp, an important source of algin used in food and medicines, is diminishing. Heavy metals and pesticide residues are found in fish in amounts above those declared by the Food and Drug Administration to be safe for human consumption.

Some species of fish-eating birds accumulate DDT in such large quantities that the pesticide is suspected of being the cause of thin eggshells and the decline of certain species, such as the brown pelican in some of their breeding grounds. Other species of commercial fish and marine birds are decreasing in numbers at an alarming rate. The specific cause is unknown in most cases. Water pollution often has been a double problem for industries; frequently, it has been necessary for the plant to condition the water before its use as well as to treat the wastewater after use. Pretreatment of the water becomes necessary to avoid a number of problems. The following has been a partial list of some of the likely contaminants and the difficulties they are likely to cause.

Alkalinity	:	Typically bicarbonates, carbonates, and various caustics.
Problems	:	Generates scale, particularly in boilers.
Hardness	:	Calcium and magnesium salts primarily. Other salts (for example, iron) are also minor contributor.

Problems	:	Scale, particularly in boilers
Sodium salts	:	Often as the sulphates, chlorides, nitrates, and bicarbonates.
Problems	:	Very bad for particular industries such as cellulose and drugs.
Silica (SiO_2)	:	
Problems	:	Scale.
Iron and Manganese	:	
Problems	:	Tend to stain. Very objectionable in the paper, textile, and tanning industries
Aluminium	:	
Problems	:	Usually not a problem for industrial purposes.
Fluorides	:	
Problems	:	Usually not a problem, except, perhaps in the production of baby foods.
Carbon dioxide	:	
Problems	:	Increases the alkalinity; enhances the corrosive behaviour of dissolved oxygen.
Oxygen	:	
Problems	:	Corrosive to iron, steel, galvanized iron, and brass.
Nitrogen	:	
Problems	:	Helps retain dissolved O_2 in water; thus increases corrosiveness.
Hydrogen sulphide	:	
Problems	:	Rotten egg odour; corosion of iron pipes, fittings, and equipment.
Methane	:	
Problems	:	Fire and explosion hazard.
Microorganisms	:	
Problems	:	Form coating in pipes; stains, tastes and odours; decompose organic substances such as cellulose.
Organic matter	:	
Problems	:	Causes tastes and odours, often absorbed by various processes; forms coloured colloidal suspensions.

The discharge water from various industries is having many of the same problems, plus some additional ones. The type of contaminant is dependent upon the particular industry and the particular process employed. It is possible to classify the contaminants into three categories:

Floating materials

Typical floating materials would be oils and greases. They have been making the water unsightly; retard aquatic plant growth by blocking the sunlight and interfering with the natural reaeration; destroy the natural vegetation along the banks; are often toxic to fish and other aquatic life; destroy water fowl; and can also be a fire hazard.

Suspended matter

A common example of suspended matter has been mineral tailings. Typically, mineral tailings form slime and sludge which smother purifying microorganisms and ruin fish spawning and breeding grounds. If the suspended matter has been organic, it would decompose using the dissolved oxygen and would produce noxious and odours.

Dissolved impurities

Typical dissolved substances would be acids, alkalies, heavy metals, and insecticides. In general, they have been making water undrinkable and destroy aquatic life. For example, phenols, even in very low concentrations (0.001 mg/l), provide a very objectionable taste and odour. They also can build and concentrate their effects through the normal food chain. For example, an unpleasant taste is noticed when, eating fish which had lived in water with a phenol concentration of only 0.0001 mg/l.

Little is known about the effect on human health of many of the common water pollutants. Nitrates have been linked to methemo-globinemia and death in infants. The ingests nitrates can be reduced by microorganisms in the digestive tract to microorganisms

nitrites : $NO_3 \xrightarrow{\text{microorganisms}} NO_2^-$. The nitrites can then oxidize the iron atom in hemoglobin from Fe^{2+} to Fe^{3+}. The result is a methemoglobin molecule, one which is incapable of O_2 transport.

Most of the other common substances probably have no major acute effects on human health. Many, however, do have chronic effects. For example, many of the organic substances have been reported to be carcinogens. Selenium causes bad teeth, gastrointestinal problems, and skin discolouration. Sodium and/or potassium are bad for people with certain health problems. On the other hand, some water impurities appear to be beneficial. A deficiency of chromium would favour atherosclerotic diseases; there is evidence that degenerative cardio-vascular disease gets reduced by a factor present in hard water but absent in soft water.

Unpolluted Vs. Polluted Water

Strictly speaking, pollution has been any departure from purity. When environmental pollution has been the topic, the term has come to mean a departure from a normal, rather than from a pure state. This has been particularly true for water. This widely distributed substance has been such a good solvent that it has been never found naturally in a completely pure state. Even in the most unpolluted geographical areas, rainwater is having dissolved CO_2, O_2, and Na and may also carry in suspension dust or other particulates picked up from the atmosphere. Surface and well waters generally contain dissolved compounds of metals like Na, Mg, Ca, and Fe.

The term hard water is used to describe water that contains appreciable amounts of such compounds. Even drinking water has been not pure in a chemical sense. Suspended solids have been removed and harmful bacteria destroyed, but many substances still remain in solution. Indeed, absolutely pure water would not be pleasant to drink, because it has been is the impurities that impart water the characteristic "taste" by which it is recognized. In light of the above facts the term *pure*, when used in a water pollution context, will mean a state of water in which no substance is present in sufficient concentration to prevent the water from being used for purposes thought of as normal. Normal areas of use include:

1. Recreation and aesthetics
2. Public water supply
3. Fish, other aquatic life, and wildlife
4. Agriculture
5. Industry

Any substance that disallows the normal use of water must be regarded as a water pollutant. Part of the complexity of the water pollution problem arises because the normal uses of water are so varied Water suited for some uses and therefore considered to be unpolluted, may have to be regarded to be polluted when other use get contemplated.

PCB's : An Example of a "Common" Dissolved Toxic Pollutant

There have been a number of particularly dangerous chemicals which have made their way into our water systems. Many of these chemicals have been in industry effluents, or at least have direct industry sources. Polychlorinated biphenyls (PCB's) are a well-known example of a toxic substance which has made its way into many of our natural water systems in this manner. PCB-contaminated rice caused

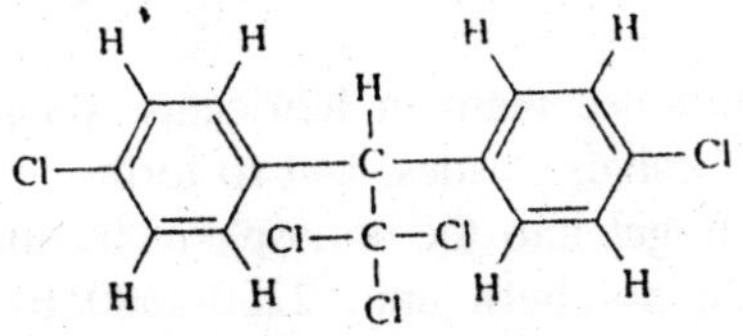

(a) DDT (single compound)

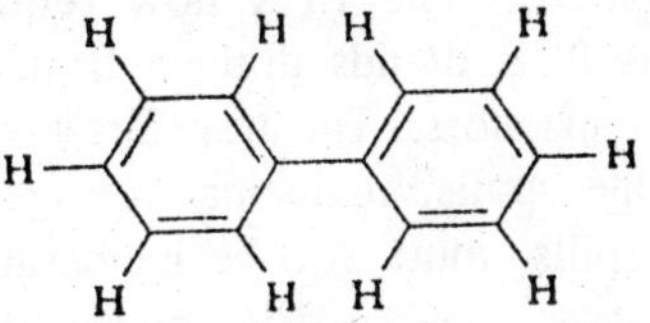

(i) Parent compound biphenyl 1. Any number of chlorine atoms can be substituted for hydrogen atoms to produce a total of 209 chlorinated biphenyls

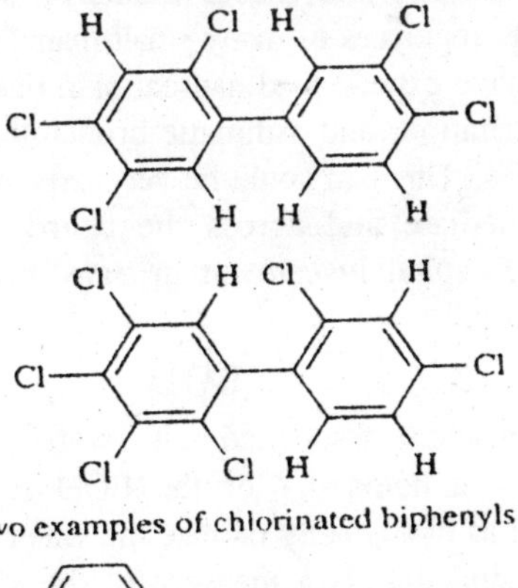

(ii) Two examples of chlorinated biphenyls

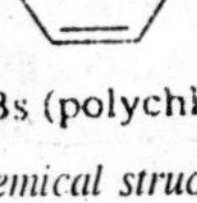

= ring of six carbon atoms

(b) PCBs (polychlorinated biphenyls)

Fig. 3.4. Chemical structure of (a) DDT and (b) PCBs.

an epidemic of a disfiguring skin disease and was blamed for 16 deaths and 2 stillbirths in Japan in 1968. Monsanto, the only U.S. company then manufacturing PCB's restricted its sales to only "closed-system" electrical applications such as transformers, where they are still used because of their thermal stability, excellent electrical insulating properties, resistance to oxidation, and low biodegradability. They have been still the safest product for this purpose, in terms of fire risk. In spite of the controls now on PCB's, they are persisting in our environment.

Other countries still use them in lubricants, duplicating paper, paints and coatings, adhesive, plastics, and so forth. The PCB's often leach from landfills, or get into the atmosphere by low-temperature incineration (PCB's do not burn until 2200-2600°F). In addition, industrial spills and other accidents appreciably contribute to the amount in our water systems. The EPA now requires high-temperature incineration of all PCB liquids drained from transformers and high and low voltage capacitors. The transformers, when drained of the PCB's, plus dredge spoils, municipal sewage sludge, and materials contaminated by spills, must also be incinerated or disposed of in a chemical landfill.

The exact health effects of PCB's are now being researched. Laboratory tests have led to liver cancer and reproductive failures in rats, chickens, minks, and rhesus monkeys. People who work with PCB's including employees of many small manufacturers and transformer rebuild shops, have experienced nausea, skin disease, fungus infections, nose and eye irritation, and asthmatic bronchitis. PCB is being banned in many countries. This ban could be very expensive for many industries if it were immediate and across the board, because of American industry's large capital investment in transformers, and the potential replacement costs.

BOD

A major problem associated with water pollution has been the "biological oxygen demand," or the BOD refers to the amount of oxygen required to biologically oxidize the water contaminants to carbon dioxide (CO_2), and thus is a measure of the suspended, colloidal, or dissolved organics. The BOD is important in that the higher the BOD, the higher the organic content of the water, and the more dissolved O_2 which will be used to decompose these organics. A lack of dissolved O_2 in the waterways kills off desirable fish (for example, trout). Though there may still be plenty of fish, there usually is a marked change in the types of fish present if the BOD is high. The presence of organic substances can decrease the dissolved oxygen in another way.

The organics, along with the nitrogen and phosphorous, primarily can serve as food for algae. These algae have been microscopic, greenish-coloured plants which live in water. In themselves they are not harmful to humans (in fact, in limited amounts, they are useful for they do create O_2 for other aquatic life by photosynthesis), but they to become aesthetically unsightly, a nuisance and a hazard to other aquatic life when they proliferate. Their growth rate has been dependent upon the conditions: they do not grow in deep fast-moving, muddy or cold waters, but flourish in lakes, ponds and slow-moving streams as long as a sufficient supply of nutrients is available. Thus, the nutrients can act as the limiting factor in their growth.

Over-fertilization with nutrients, or eutrophication, can take place naturally. The algae eventually die and decompose. In addition to producing a slimy scum and obnoxious odour, they required O_2 and hence can drastically decrease the dissolved O_2 supply as they decay. The particular nutrient which has been in limited supply acts as the "limiting factor" in algae growth and multiplication.

Usually it has been very difficult, if not impossible, to control the nitrogen. Farm runoff supplies much to the lakes and streams. Some algae have been even nitrogen fixing, and therefore can obtain all the nitrogen they wish directly from the atmosphere. Organics could be controlled somewhat, but usually the phosphorus is targeted as the nutrient to try to minimize (hence the emphasis on low-phosphorus detergents). Manufacturing produces large quantities of BOD-20-25 million lb annually. The major contributors have been the chemical industry (about 44% of the manufacturing BOD), the pulp and paper industry (about 27%), and the food processing industries (about 20%). Table 3.4 lists typical BOD and suspended solids levels produced by a variety of industries.

Related to the BOD, and often measured instead has been the COD, the chemical oxygen demand. The COD is a measure of the amount of oxygen required to *chemically* oxidize the contaminants to CO_2. The value of the COD has been higher than the BOD, for strong oxidants are used which force many substances to react which wouldn't react using biological microorganisms. The COD value thus represents almost 100 percent of the total organics present. Dissolved sulphites and ferrous compounds, which can act as reducing agents, can also help deplete the O_2 supply.

Table 3.4. Typical Process Waste Strengths

Type of Industry	*Process Wastes (mg/L)*	
	BOD	*Suspended Solids*
Petroleum refining	310	370
Leather and leather products	1300	1700
Flat glass	310	370
Concrete, gypsum, and plastic products	310	10000
Primary metal industries	310	370
Industrial launderers	870	1100
Meat products	800	520
Dairy products	2100	370
Preserved fruits and vegetables	1400	370
Grain mill products	3200	4500
Barkery, mixed	2500	1103
Fast and oils	1300	1100
Malt beverages	2100	980
Malt, wine, brandy, distilled liquor	650	370
Seofood and fish	1400	600
Food preparations	4100	930
Medicinals and botanicals, fermentation	4500	4700
Paints and allied products	2500	600
Industrial organic chemicals	310	370
Miscellaneous chemical products	2500	2600

ACID WATER SYSTEM

Another problem often experienced in natural water systems due to industrial pollutants is a change in the pH of the water. Natural waters now vary in pH from 4.0 to about 9.5. Some of the acidification of the waters occurs naturally. Carbonic acid, formed by the reaction of CO_2 and water in the atmosphere, is picked up in the normal precipitation process, so that rainwater is typically at about a pH of 5.6. However, acid mine seepage and, particularly, acid rain, have had a major effect on the pH of many lakes and streams. Acid rain, for example, can have a pH as low as 3.9. In areas where calcium or lime are present in the rocks or soil, some of this additional acid is neutralized. The pH of the water has a major effect on the type of fish and other aquatic species present.

A recent study revealed that pH was 30 percent more important than water temperature and 50 percent more important than dissolved

O_2, at least in the case of bass populations. Both trout and bass prefer alkaline streams, though that may be due somewhat to the additional advantage that more food is usually present. Alkaline streams are usually found in localities where the rock and soil contain $CaCO_3$. Rain, with its normally small amount of carbonic acid (H_2CO_3), dissolves the $CaCO_3$ present and forms calcium bicarbonate. The calcium bicarbonate is a source of CO_2 for photosynthesis by green plants.

In addition, the $CaCO_3$ itself is necessary for the growth of crustaceans such as crayfish, scuds, and sow bugs, another source of food for fish. Trout are able to withstand somewhat acidic waters; in fact, pH values as low as about 5.0 have been possible. However, in these conditions their growth is poor. This effect on even fairly remote, high elevation lakes has been a serious problem of particular concern to the sports fisherman. Work is being undertaken by several groups to develop a means of neutralizing some of these acid lakes. It has been experimented with dumping agricultural crushed lime by helicopter into selected lakes, and has found it feasible on a cost/benefit basis. Studies are also underway to develop an acid-resistant strain brook trout. The situation does not look as if it will improve, however, in the near future.

Other Considerations

In general, industrial manufacturing is the leading source of controllable man-made water pollutants. In addition to the large quantities produced, industrial wastes are also much more likely than other types of waste to contain chemicals which are not biodegradable, toxic organic substances, and trace elements which must be removed. When one considers water pollution control and management, it is necessary to consider, in addition to the nature of the contaminants, also how those substances travel through the water systems. As they travel, the contaminants may be degraded by chemical, physical, or biological processes. They migrate by flow or convection, disperse throughout the water systems by diffusion and/or mixing.

In order to make the best decisions as to how to handle these waste waters, it becomes necessary to consider all of these processes and the overall transport. Only in this way can wise decisions be made as to which pollutants must be removed, and, which could remain in the effluent. Computer programmes have been developed for this purpose, and have been taken advantage of by various industries. The size of pollutant particles also affects their behaviour. The sizes may be categorized as follows:

1. Settleable	> 100 μm	largely organics
2. Supracolloidal	1-100 μm	
3. Colloidal	1 nm-1 μm	bacteria, viruses
4. Soluble	< 1 nm	inorganics

The larger particles can be able to absorb and bind various substances on their surfaces, and thus can act as (for example) sites to promote bacterial growth. Clays can absorb dissolved organics such as pesticides. These clays can then settle to the bottom of the lake or stream. In this way the pesticides and other organics become incorporated into the sediments. The release of these pollutants back into the body of the water has been unclear, and may pose a serious problem in the clean up of our now polluted waters. The intentional use of soil systems to strain and absorb wastes and to remove even bacteria and viruses has had some limited industrial applications. Often, the soil systems can degrade undesirable compounds to stable ions normally found in ground water—for instance, to nitrates, phosphates, carbonates, and sulphates. On the other hand, sometimes chromate ions, phenols, and many other "toxic" substances have been not degraded, and can migrate underground many miles, contaminating water supplies far from the pollutant source.

The Costs of Water Pollution

The costs of water pollution have been probably even more difficult to estimate than the costs of air pollution. Who can measure the cost of not being able to swim in some river or not being able to catch fish in it? Who will be able to measure the aesthetic cost of a polluted lake or waterway? It is even valid to try to express these costs in monetary terms? It is evident that for a very modest price people are able to enjoy a plentiful and generally pure supply of drinking water. The cost has been certainly small in comparison with the cost that would be incurred by society by repetitions of the waterborne epidemics of the recent past.

Ganga Action Plan

The Rs. 560 million Ganga Project that started on a war footing last June has been aimed at setting up sewage treatment plants to tackle 1000 million lines of sewage per day from 27 cities on the bank of river before it being discharged into the river. It would also renovate and augment considerably the capacity of existing sewage pumping stations, construct new pumping stations, and lay additional sewers, Mr. S.S. Bagga, the Divisional Commissioner and Incharge of the project said. Arrangements are being made to control the discharge of industrial effluents, which are the major sources of pollution besides sewage. Ganga Action Plan covers 27 cities along the river. The work is going in most of the towns.

In Varanasi, under a multisectoral city Development Package, ghats are being renovated; sewers are being intercepted and work has been in full swing on two of the three sewage treatment plants. In Kanpur, the sewer network and treatment plants at Jajmau have been part of larger package of development schemes. In Allahabad Gaughat pumping station is being expanded. Nallahs discharging waste into Yamuna river are being intercepted, work is also proceeding fast at the Naini Sewage farm. The major aspects of the programme have been interception and diversion of sewage now flowing into the river, integrated sewage treatment plants with energy recovery, low cost sanitation, etc. A wide range of schemes are working to deal with domestic wastes and industrial effluents polluting the river.

The Ganga Project Directorate monitors closely all major schemes costing Rs. 50 lakhs or more. State level specialised agencies have been made responsible for execution and maintenance. Many schemes are expected to be completed by the end of the Seventh Plan. The recovery of resources from the treatment of wastes, monitoring of water quality and conservation of the fish and aquatic life in the river have been important aspects of Ganga Action Plan. All the major universities along the river and scientific organisation under the council of Scientific and Industrial Research and ICAR network have been involved in the formulation of the schemes. Arrangements are done for regularly monitoring water quality of the river at various locations and at different times. The main thrust of the Ganga Action Plan has been to prevent pollution so that the purity of the river can be restored all along its course.

4

Biotechnological Detection of Pollution

General Bioassay in Pollution Monitoring

Efforts to monitor pollution through biosystem approach have gained importance in recent years. Environmental protection agencies realize that new alternate technology holds tremendous potential in restoration of environmental quality and its management. Screening of xenobiotics in environment with living test systems in one such side. No water pollution are complete until biological aspects are included. Bioassay is not a tool, by important rule to allow/disallow the discharge of a toxic substance.

The availability of techniques help toxicologists and scientist to monitor environmental health from diagnostic, preventive and remedial point of view. Till now, conventional surveillance was largely based on physical and chemical methodologies. But in any pollution study, knowledge of mechanism of biological response play important role as ultimate damage is effected on living system. It becomes often difficult to detect the intervals and chronic pollution effects through conventional technologies. This is due to damage to biological systems may be brought about a very low concentration of the pollutant, much below so called *permissible limited* (MATC = Max. allowable toxicant concentration) as recommended. These hidden injuries often be resolved by biosystem approach. Cost-effectivity is an advantage in long run, when knowledge of 'Ecological Equivalent' may offer possibility of extrapolation of data from one to other region. Bioassays don't require assumptions and they provide the means of quickly screening samples for toxicity assessment. Bio-monitoring serve in a complementary fashion with other strategies in defining toxicity outset.

In this regard, many groups of plants, microbes and animal test samples have been used at different organization levels. This chapter tries to give some techniques developed so far to trace pollution-induced responses and the possibility of their use in monitoring programs.

Choice of Criteria or Parameters

The parameters are crucial in bioassay studies as to reliability, predictivity and universal acceptance. This is more seeming when taken into account the possibility of quality variability in living systems coupled with influence of environmental factors upon them. However, the techniques of bioassay largely has 3 types of criteria namely, symptoms of visual damage, genotoxicity test and assessment of the responses at the metabolic levels.

Visual rating

At the beginning stage, it was largely based on visual rating only. Hence the concept of LD_{50} or LC_{50} emerged for fishes, i.e., the dose at which 50% of test organism is affected. To determine this computer programs also been developed, and the growth rate are considered. For microbial growth, turbidometric analyses are used more often now a days. In case of higher seed viability, germination frequency, growth rate of different parts and any damage symptom of leaves and other parts are considered. Presence or absence of a particular species in a polluted environment (*indicator*) is another good criterion, and however no single indicator species can fulfill all the purposes equally well, hence multi-species testing concept has emerged.

Genotoxity test

This involves assessment of damage at cellular and sub-cellular levels. Cells are primary sites of interaction between chemicals and biological systems. The importance of toxicity is based on assumption that cellular changes are finally reflected in the metabolic and morphological disorders at the organismic, species or population levels. The parameters at cellular level include different biomolecules, organelles, immune reactions and membrane processes. These involve simple and inexpensive tests.

Cytotoxic tests depend on the estimation of chromosome damage including, breakages, sister chromatid exchange or micronuclei counting. Dis-function of plasma membrane is caused by lipid largest proportion of oxygen affecting its fluidity and structural integrity, can be measured by conductivity studies indicating ion leakage or enzyme leakage like LDH. While lysosomal viability is rated through the neutral red dye retention assay. Lately DNA probe is also being considered as a natural consequence to these parameters, particularly helpful in identification of Pathogens in water samples.

Metabolic rating

With knowledge of physiological and biochemical mechanisms of metabolic, processes, it was thought worthwhile to consider the changes in some of these parameters induced by pollution stress as unquestionable 'biomarkers' and help in visualizing the threatening danger.

Biomonitoring involves both qualitative and quantitative assessment of these markers. They include changes in contents of chlorophyll molecule and its fluorescence kinetics, contents of soluble protein, nucleic acids, changes in activities of key enzymes, their reaction kinetics and electrophoretic assessment of isozymes or the induction of some stress proteins.

Examples Highlighting the Utility of Bioassay

Plant test systems

Among test in plants, *algal bioassay* have been used since 1970s and is due to their high sensitivity, easy availability and cultural facilities in limited areas and are good indicators of pollution. The primary criteria considered are growth rate, productivity and LC_{50} values. At the metabolic level, photosynthetic efficiency has taken into account using C^{14} assimilation, O_2 evolution and by noting degradation of *in vivo* chlorophyll fluorescence. Action of some key enzymes has been considered as well.

Some of algae, commonly considered as test materials, are *Chlorella*, *Scenedesmus*, *Selenastrum*, *Navicula*, *Spirulina*, *Anabaena* and *Microcystis* among the microalgae, while among the marine Macro algae *Ulva*, *Fucus*, *Laminaria*, *Macrocystis* and *Codium* are worth mentioning.

Experiment of algal toxicity is been reviewed by many. For quantifying the responses LC_{50}, Cu-equivalent approach and (TU = $100/LC_{50}$) concepts been used. Palmer's (1960) *pollution index* helps composite rating of organic pollution. In case of marine macroalgae, changes in frond structure and biomass have been used as pollution detection criteria. Benthic community diversity has also been considered.

Algal species are often used as bio-indicators of pollution. Dominanace of Cyanophyceae in the phytoplankton community of a water body indicates its eutrophic nature. *Microcystis* causes 'waterbloom' by producing reddish waste of luxuriant growth in aquatic systems due to availability of organic phosphates in abundance. *Navicula*, on the other hand, indicates metal pollution in lakes, being the dominant community where metals like Pb. Zn and Cu are present from mine effluents.

Bacterial bioassay is used since long for detection of faecal pollution in potable water through 'coliform' test. At present, obligate anaerobic bacteria and their viruses are considered to be more suitable indicators of water pollution, while the bacterium *Salmonella* been used in *Ames Test* to find atmospheric mutagens with help of its different mutant strains. Bacteria are good indicators of organic pollution.

Phenomenon of bacterial *bioluminescence* been used as indicator in analysis of atmospheric gases and other compounds like So_2, formaldehyde, ethyl acetate etc. Bioluminescence is an enzymatically catalyzed light-emitting reaction in living cells. *Photobacterium phosphoreum* colonies used in a special photodetector, where change in emission of light due to pollutant effect is found by a sensor, then made bigger and recorded in a computer. These photo-bacteria also be used for investigating pollution potential of waste water including that of marine ecosystems.

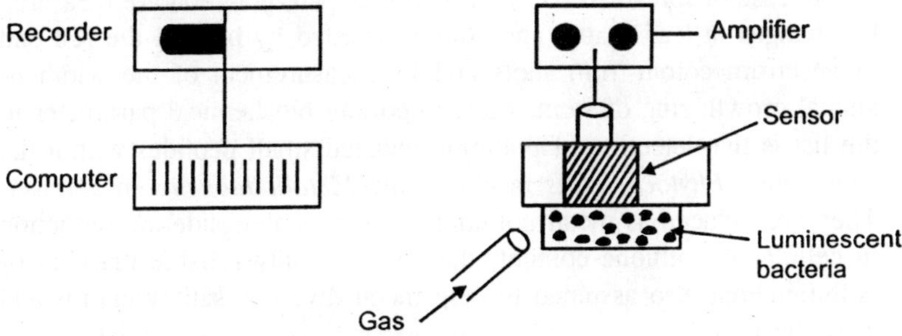

Fig. 4.1. Schematic representation of a photodetector.

For 'Mapping' cities for atmospheric SO_2 detection, Lichens are used. They grow less in areas where SO_2 concentration in air is relatively high. Lichens serves as a litmus to atmospheric gas pollutions.

A number of forest and aquatic mosses such as *Bryum*, *Stereophyllum*, *Frontinalis* and others have been used to monitor metal pollution in rivers by measuring their shoot up growth, while *Sphagnum* has been used to monitor atmospheric deposition of As and Se in Canada in the neighbourhood of mining and extracting metal from ore areas. Aquatic ferns like *Azolla* and *Salvinia* shown promise in metal pollution monitoring.

Aquatic plants more frequently used for wastewater monitoring among vascular *macrophytes*. Many crop plants and road side trees helped monitoring atmospheric deposition of pollutants. Water hyacinth (*Eichornia crassipes*) and duck weed (*Lemna minor*) have widely used to monitor aquatic metal pollution. Other aquatic plants that have been found useful in this regard are *Pistia stratiotes* another duckweed *Spirodela*, *Vallisneria* and *Hydrilla*. In all these physiological and biochemical parameters like soluble protein, nucleic acid and enzymes like peroxidase and catalase, besides chlorophyll estimation, have proved to be excellent bioassay parameters. Photosynthetic and respiratory activities also serve as important criteria. Incase of peroxidase activity, it has been noticed that there occurs a sudden flow when subjected to metal pollution. So it serve as a 'biomarker' of pollution, while mercury inhibits nitrate reductase activity. Other key enzymes which were tried in test system include polyphenol, oxidase, dehydrogenases and enterases. Mukherji (1990) estimated a number of physiological parameters in road-side weeds like *Calotropis*, *Lantana*, *Solanum* and *Cronton* to assess Calcutta city air pollutant effects and clearly noted that plants away from main road have less disturbances in their metabolic activities.

In case of forest trees having long-life span, pollution are measured by changes in leaf shape, and colour is tested by beyond the red end of spectrum colour film shots and by measurement of the width of annual growth ring of stem. Most important biochemical parameter to the list is the detection of pollution-induced small peptides within the cells called *Phytochelatins* in plants and *Metallothioneins* in animals. They are induced as metal pollution responses alongside the reduction in cellular glutathione content, they are not only reliable markers of pollution, but also assumed to help metal divert to satisfy claims and detoxification.

For monitoring ecotoxicity at population level, using demographic techniques where parameters like biomass turnover, survival and phenology are considered, organisms with short life span are more desirable, for this among plants algal system for microbial toxicity testing and *Arabidiopsis thaliana* for higher plant toxicity have been used.

Animal test systems

Among the animals of a region, toxicity to fishes have been considered since 1950s as a good criterion of xenobiotic bioassay to provide basis for strategy formulation in the release of toxicants. It has provided the basic tool in monitoring pesticides, phenolic compounds and wastes from tanneries and textile industries. The concept of LD_{50} first emerged from studies on fishes. Toxicity assessment has been done on morphology, behaviour, changes in different organs, muscles and metabolism as well. Inhibitory effects of xenobiotics on fishes have been investigated on enzyme AChE, which is a neurotransmitter and has been found to be a good marker of pesticide pollution. Some of the common fish types in bioassays are *Catla*, *Labeo*, *Channa*, *Teleost* and *Tillapia*.

Other faunal groups include *Protozoa*, particularly the ciliated ones, as good bioindicators through changes in their behavioural patters, and thus the concept of *ethogram* has developed. A number of zooplanktons, like phytoplanktons have helped much in evaluating toxicant effects in aquatic systems. They include the well-known *Daphnia monga* and other crustacean species. They also accumulate and concentrate a variety of harmful materials including heavy metals, qualifying them as effective, indicators of metal pollution in aquatic environment.

Rotifers are a kind of helminths among the zooplanktons and grow mainly among aquatic vegetarian. They form another class of fauna which have proved to be good indicators of saprobity and trophic levels of water indicating the contents of putrid organic matter as given by BOD_5. The genus *Brachionus* is connected with eutrophic water. The quotient QBT has been considered to be a good index of trophic levels, viz., oligotrophic when QBT is <1, mesotrophic 1-2 and eutrophic >2. However, all attempts are still based on lab tests and yet to be included in field bioassay, the advantage of applying rotifers in biomonitoring of water lies in their easy recognition, easy cultivation, slow growth rate and year-round availability.

Although environmental biotechnology is still in 'formation stage', there is no question now with respect to its use as pollution monitoring parameters along with the conventional techniques. It is hoped that when some of test systems and the parameters are standardized, pollution biomonitoring will become more sophisticated and will be place on a sound footing. But a single bio-test may not always satisfy the criteria and which a battery of tests and multi-species system may be desirable. However, the standard test organism should meet certain criteria, which are as follows :

1. It should be sensitive to a pollutant.
2. It should be able to readily absorb/adsorb the toxicant.
3. It should be readily available throughout the year and is of common occurrence.
4. It should have features which are directly measurable and predictive.
5. It should be useful in providing inexpensive or cost-effective measure.

Cell biology in Environmental Monitoring

This is special branch in biology encompassing the structural and functional aspects of cells and their components. Developments in allied sciences like bio-chemistry, biophysics, genetics and molecular biology coupled with instrumentation have widened the scope of cell biology, is being effectively used as an important parameter of screening environmental mutagens and carcinogens.

Cell Biological Methods

Cells are the sites of primary interaction between toxicants and biological systems and cell biological methods act as 'biological dosimeters' at cellular level damages.

The tests under the orbit of cell biology aim to trace the damages caused by different environmental toxicants on different cellular parameters like membrane, cell organelles and chromosomes, carrying the genetic materials like nucleic acids and proteins, they also help in understanding the basic mechanism of toxicity, mutagenecity and carcinogenecity.

The use of this aspect of study has assumed such an important that a separate discipline called Environmental Mutagenesis has been instituted. It got a major thrust with the exponential growth of industrial chemical production and realization of their effects on human genetic system. Like any other scientific discipline, mere identification of phenomenon led to the technological innovations of detection at both in vitro and in vivo levels to find remedial measures, one such being the possibility of amelioration of radiation damage in human genetic systems to a great extent by cysteamine injection. It forms disulphide (–s–s–) bridge with cellular protein and offers protection. If the source of damage is known then test may help quantitation and prediction of further consequences and act as warning systems simple and not expensive.

Test Systems for Assessment of Genetic Damage

Since 1940, various test system have been developed using microorganisms, plants, lab animals and human cell cultures to assess the toxic and mutagenic potency of numerous chemicals and radiations and even 'living mutagens' or virus and bacteria induced mutations.

Ames test

The test based on cell-biological principles include the classical Ames test involves the use of series of strains of the bacterium *Salmonella typhimurium* with specific ability to detect certain types of chemical mutagens. It is a very widely used preliminary short-term testing and screening of drugs, agricultural chemicals, cosmetics, food additives, inorganic metals and other pollutants, is based on principle of the frequency of reversion of mutations at the histamine locus been used for such type of detections.

Cytogenetic assay

In plants and animals detection of permanent genetic damages takes longer time depending on life span. Thus in these cases, chemical induced disruptions in the chromosome system, be it in vegetative or reproductive cell, is taken ass the criterion of assessment of mutagenecity or carcinogenecity as they constitute the basic hereditary materials. However, insects with shorter life cycle (*Drosophila*) or '*Arabidiopsis*' system among plants permit the analysis of evens through the entire life.

Plant systems have helped reliable screening and monitoring hazardous environmental chemicals are pea, maize, soyabean, *Crepis* and *Tradescantia* and this has unique features which provide 2 ideal genetic end points for mutagenecity testing, namely 'Stamen Hair' assay and micronucleus test. The stamen hairs are highly sensitive to gaseous chemicals like Freon-22, Benzene, Ozone, NO_2, SO_2 and serve as indicators by changing their colour.

Chromosome damage

The most common method of toxicity test of chemicals is called '*Allium test*' to check the extent of induction of chromosome aberrations in form and behaviour and may be of different types including fragmentation of chromosomes, bridge formation and disruption of normal cell division serve as a clear 'hazard indicators' and can be quantified.

The severity of damage may depend on the chemical make up of the toxicant. Incase of phenolic compounds, it has been observed that the toxicity is influenced by the number and position of —OH groups.

Genotoxic potency wise they can be put as Pyrogallol > Resorcinol > Gallic acid > Phenol > Guaiacol.

Micronucleus test (MNT)

This is another method where very small nuclei are stored within the cell formed by large-scale fragmentation of chromosomes. The extent of such nuclei formation directly co-related to the severity of damage. This is now routinely used for screening mutagenic compounds due to relative simplicity and rapidity and is most practical for analysis of bone-morrow cells of animals *in vivo* after exposure to chemicals.

Sister-chromatid exchange

This is an effective parameter of cytogenetic assay is based on damage caused to DNA molecule and mis-exchange of chromosome segments. This is characteristically revealed first by Bromodeoxyridine (Brd-U) dye and then stained with a fluorescent dye technique. The potentiality of this test is that even after a short exposure, it can be easily detected.

Membrane damage

Apart from above nuclear parameters, damages to cell membrane and cellular organelles be found in a number of ways.

The plasma membrane is first structure encountered by a toxic agent upon reaching the cell. The membranes are sites of uptake, deposition and elimination of chemicals. The lipo-protein composition of the membrane act as selective barrier due to its semi-permeable nature. Many compounds can change the membrane structure and its fluidity by the 'lipid peroxidation', i.e., oxidative destruction of the membrane. Dysfunction can lead to cell and organismic death. Inside cells, other membrane systems include mitochondria, nucleus, vocuoles, lysosomes, etc., act as effective sites of compartmentalization.

The loss of semipermeability of membrane be measured by leakage of enzyme like LDH or efflux of electrolytes or through the uptake of *trypan blue*. Break of Lysosome can also be assayed by *Neutral red* retention test i.e., loss of capacity of lysosome to retain this cationic dye after being damaged and been used a *biomarker* of cell viability. The physical damage to different membrane or deposition of toxicant granules in different cellular compartments can be detected through light, electron and phase contract microscopy including imaging techniques.

Developments in techniques of plant and animal tissue culture helped in monitoring these cellular damages in successive cell cycles.

Human lymphocyte culture is now a valuable tool for routine monitor of persons exposed to occupational hazards of chemical substances.

Dangerous chemicals like vinyl chloride, methyl-isocynate or phosgene of 'Bhopal Gas' fame can cause total disruption of cellular components leading to death. Thus monitoring of environmental toxicants through cell biological methods gives better insight into the problem and may help development of remedial measures. More over long term exposure to very low concentrations of toxicants may often be revealed only through the detection of damage to cellular components vis-à-vis the genetic components.

Molecular Biology in Environmental Monitoring

Molecular probe is based on small segment of DNA or RNA capable of recognizing complementary base sequence in target DNA from sample organism in monitoring of environment, and also through sequence recognition of specific protein as is done in *immunoassay*. DNA probe or DNA-gene probe is known, helps detection of gene rather than its product. While immunoassay identifies the amplified gene product.

The developed molecular probes are used in various biotech studies. In the arena of environmental analysis, these are useful in diagnosis of infectious disease, identification of water borne pathogens, microbes in mixed cultures and food contaminants. They are very sensitive, accurate and faster too. While the conventional tests, based on morphology and physiology, may take few days (4-5), the molecular technique takes only a few hours.

Initially it requires time and effort to develop a particular assay. But once optimized, it offers tremendous advantage over the traditional analytical technique.

DNA-Gene Probe

This is used variously in basic and applied aspects of bio-technological studies. The principle likes in the 'hybridization' of complementary base sequence of DNA strands, where a given strand can pair with its complimentary one.

Step 1 : Extraction, separation and isolation of DNA fragments.

Step 2 : The double stranded DNA is heated and dissociated into 2 molecules known genetic function constitute the probes.

Step 3 : Labelling of probe : For easy identification of such a probe, segments of DNA are 'labelled'. These labelling could be achieved either with (1) radioactive isotopes (32_p or 35_s) or with (2) non-radioactive methods using biotin (vitamin H) or fluorescent technique at a particular wave length.

Step 4 : DNA Hybridization : If target DNA in the test sample has matching base sequence with the probe, then a double helix well be formed. Such double strand constitutes the "hybrid" molecule and the process is known as DNA hybridization. The hybrid molecule can be identified easily due to labelling of probe, and provide genetic information of the microbe in question and can be conducted on mixed cultures as well.

Step 5 : PCR Technique for DNA amplification : DNA probes mostly need 10^5 to 10^6 copies of the target gene to produce a reliable positive signal and thus need to be amplified. This amplification is done with help of a thermostable enzyme DNA polymerase and can create millions of copies of DNA from a single copy of the target DNA molecule. This process is known as *polymerase chain reaction*. This invention now been undoubtedly established as one of the most far reaching developments in molecular biology and has become a tool in hands of biotechnologists. The technique has been justifiably compared with 'finding a needle in a haystack and then producing a haystack of needles by way of amplification'.

The PCR method involves 3 steps, denaturation, annealing and extention of DNA. These steps continue a cycle and with 25-30 cycles, a massive amount of DNA can be obtained. Infact, PCR machine has been developed which allows the operation to be completed within few hours. It requires the presence of the enzyme and 2 synthetic *primers*,

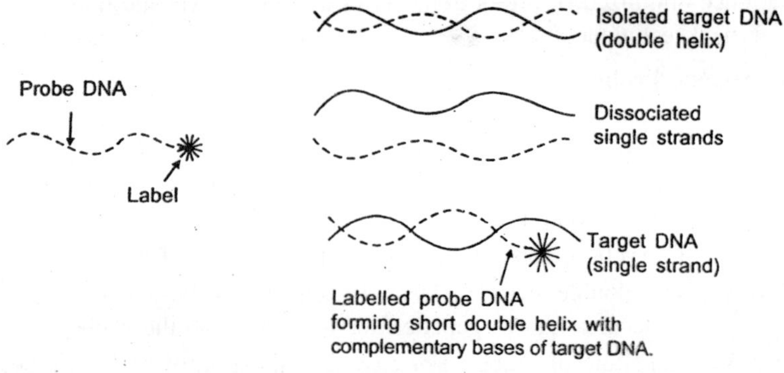

Fig. 4.2. DNA probe.

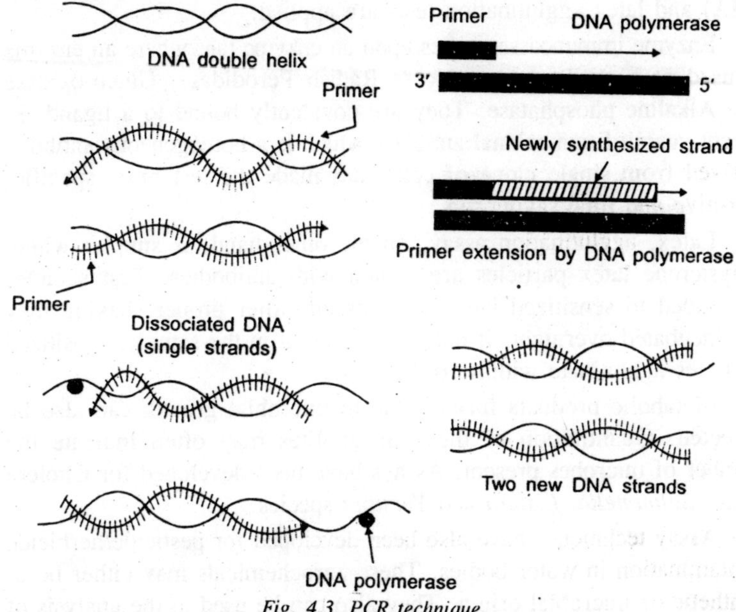

Fig. 4.3. PCR technique.

which are complementary and bind the 2 sites on either side of target. DNA to be amplified and then extend to give a new strand. When these 2 ingredients are supplied, PCR can literally amplify almost any piece of DNA.

Using both DNA probe and PCR technique it is now possible to detect bacteria and viruses for water quality monitoring upto 1/100 ml. approx. Bej et al. (1990) could identify coliform bacteria and enteric pathogens like *Shigella* and *Salmonella* in water through use of lac Z lam B gene analysis.

It is thus possible to quickly assess the bacterial safety of potable water supplies in conjunction with conventional techniques. It has also been developed for food borne pathogens like *Listera* and *Vibrio Cholerae*.

Immunoassay

This techniques based on antigen-antibody reactions are widely adopted in various environmental analysis including diverse pesticides/ herbicides, and identification of microbial pathogens and their toxins in food and water. They are useful only in those where substances show immunological properties. Overview of these techniques have been made by Swaminathan and Konger (1986), Vanderlon et al. (1988)

and Notermans (1990). Of these techniques, both enzyme immunoassay (EIA) and latex agglutination tests are applied.

Enzyme immunoassay relies upon an enzyme tag, where an enzyme is used as a marker, viz., Horse Radish Perodidase, Gluco-oxidase and Alkaline phosphatase. They are covalently bound to a ligand. In recent years of monoclonal antibody which is a homogeneous antibody derived from single clone of cells, has made the test more specific, sensitive and time saving too.

Latex, agglutination assay, on the other hand, is simpler where polysterene latex particles are coated with antibodies. Test samples are added to sensitized latex particles and after proper shaking they are incubated overnight. If antigen is present in the sample, sensitized particles agglutinate within 10 minutes.

Metabolic products formed due to microbial growth can also be detected. Quantitation of these metabolites may often indicate the number of microbes present. Assays have been developed for Cholera toxin, *Salmonella*, *Listera* and *Yersinia* species.

Assay techniques have also been developed for pesticide/herbicide contamination in water bodies. These agrochemicals may either be of synthetic or microbial origin. They also can be used in the analysis of phytosanitary quality of seed products for scale as disease-free ones.

Some of the pesticides which have been measured by immunoassays are triazines, aldrin, 2, 4, D, DDT, paraquat, glyphosate, etc. Moreover, these methods can detect pesticide concentration in water much lower than the detection limits of conventional systems.

Immunoassays for environmental contaminants are attractive for their cost-effictivity, sensitivity and accuracy. In approximately 80% of cases the results obtained through them agree with the conventional techniques.

Bioluminescent "Lux Reporter Gene" Technology

It is a new concept on plasmid encoded certain gene sequences that produce assayable flurogenic signals when genes for catabolism are expressed in luminiscent bacteria *Vibrio* or *Photobacterium*.

It is a highly sensitive and attractive way of monitoring contamination of ground water and subsurface soil. The lux reporter genes produce visible light in host bacterium and some bacterial strains have developed through gene cloning in *Pseudomonas* for tracing the degradation of Napthalene, Salicylate, Toluene and Xylene. The changes in emission spectra occur due to changes in fluorescent protein and the enzyme luciferase. In future, fibre-optic methods of remote light sensing and cell-immobilization technique would help in online monitoring system.

Biosensor in Environmental Analysis

This is an analytical device based on sensing response of biological materials in combination with an electro-chemical transducer to convert the biological reaction into a digital electronic signal which is proportional to concentrate of target. It in highly specific and accurate in detection.

The concept of biosensor is attributed to Clark and Lyons (1962). The 'Enzyme Electrode' was christened by Updike and Hicks (1971). Since then, it gained popularity and array of sensors been developed based on different bio-detecting elements. Till recently, it has widely used in monitor of efficiency of industrial bio-processes, health care for detecting blood glucose level, urea in body fluid and in military purposes for monitor of toxic gases in warfare. They are now finding more application in the realms of environmental monitoring and pollution detection. A biosensor comprises 3 components.

(a) Biological component consists of immobilized biological sensing materials, namely., enzymes, immune-agents, lectin, DNA, whole microbial cell and higher plant or animal tissue slices. The advantage being conferred by biological material is the ability to operate at surrounding temperature.

(b) Physical component is a transducing element which converts biochemical interactions into electrical or optical signal and could be made of platinum, gold or graphite.

(c) Interface is a polymeric film or membrane linking the biological component to the tranducer.

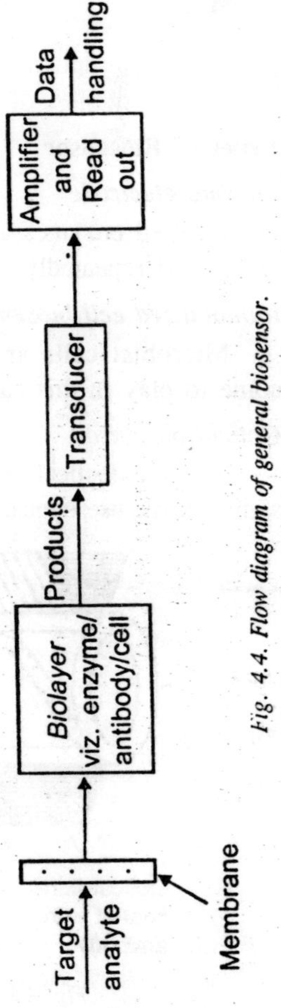

Fig. 4.4. Flow diagram of general biosensor.

Sensors could be of indirect or direct type. In former two electrodes are used, and in latter, the biological component is placed on the electrode offering more sensitivity and accuracy as electrons are transferred from analyte to enzyme or other sensing element, and directly to electrode.

Sensors, whether enzyme or whole cell, mainly work on principle of gas electrodes, viz., O_2, CO_2, H_2O, NH_3, H_2S, etc.

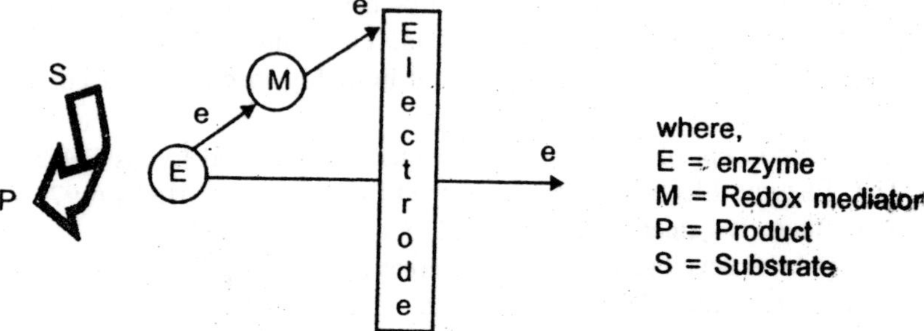

Types of Biosensor

Enzyme electrode

Purified enzymes are required and if immobolized electrode can be utilized repeatedly.

Immobilized cell biosensor

Microbial cells are used. As a multiple enzymes of cell may come to play in this case, care should be taken about selectivity.

Optical biosensor

This uses optical principle to convert biological activity into electric signal. Some use optical fibres and fluorescent molecule tagged antibody

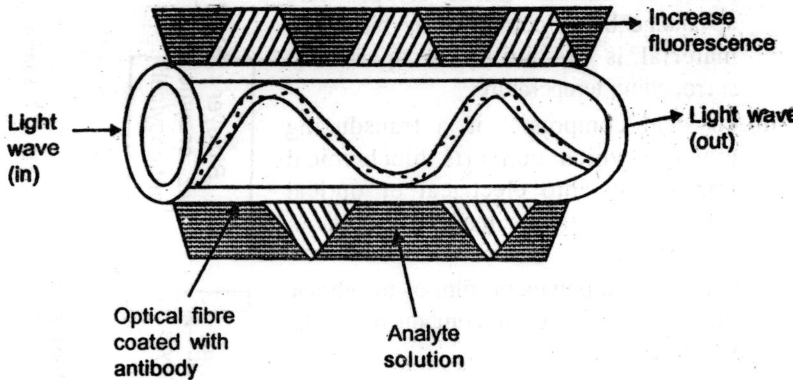

Fig. 4.5. Schematic diagram of optical sensor.

to find antigen. Since these are non-electrical, they have advantages of *in vivo* applications and multiple analyte can be examined using different monitoring wavelengths. The fluorescence intensity is measured by a detector and converted into electrical signal. They are of immense value in medical science. They can be used in environmental studies. Optical sensors are resistant to electrical disturbances, have short response time and wide range of materials be used for their fabrication. Other advantage is that the signal remains unaffected even if 'fouling' occurs on membrane surface from bio-film. The main drawbacks of this sensor are its fragility, cost and bulkyness. However, recent development in fibre optics has helped the system in many ways.

Ion sensitive field effect transistor

It is a semi-conducting device for measuring extremely small changes in analyte. Biosensor for environmental application does not differ in principle from those used for other purposes. Kits have been developed to identify specific pollutants. Biochips can monitor change in environment, viz., microbial response to presence of heavy metals and organic toxins in soil, water and air.

H_2O_2 biosensor

It is based on oxidoreductase enzyme, where O_2 is used as H_2 acceptor forming H_2O_2. The rate of production of H_2O_2 gives the measure of analyte present in sample.

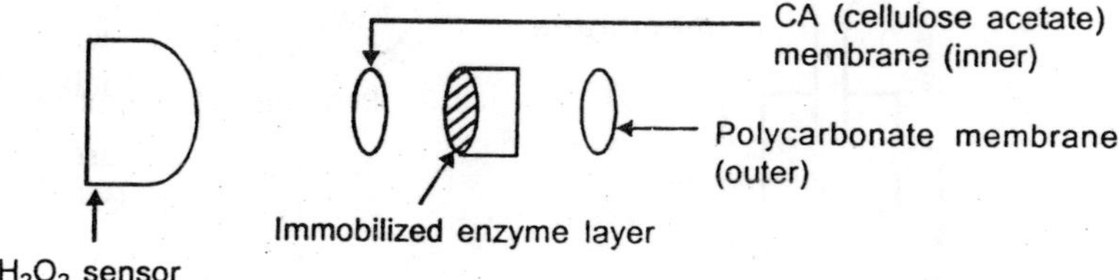

Fig. 4.6. Schematic diagram of H_2O_2 biosensor.

The membrane could be of cellulose acetate or collagen on which the enzyme is immobilized. The enzymatic membrane can be replaced as and when necessary. H_2O_2 sensors are used in continuous monitoring of blood glucose in vivo, different industrial process control and waste water treatment by placing the sensor in flowing stream.

Microbial biosensor

Composed of immobilized microbial cells and be classified as either (1) Respiratory activity measurement type or (2) Electro-chemically active metabolite measurement type.

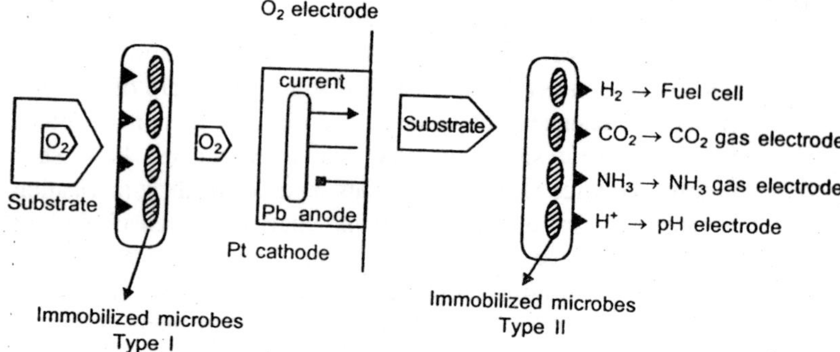

Fig. 4.7. Microbial sensors.

Besides there could be thermistors for measuring heat generated due to metabolic activities with luminescent bacteria where change in light intensity be detected, viz., decrease in intensity due to metabolic inhibition by heavy metals or other toxins. For microbial sensor, usually membrane entrapment of cells technique is used these days, rather than the get type. After the membrane is prepared, it is stored in a buffer solution ((pH 7) at 4°C.

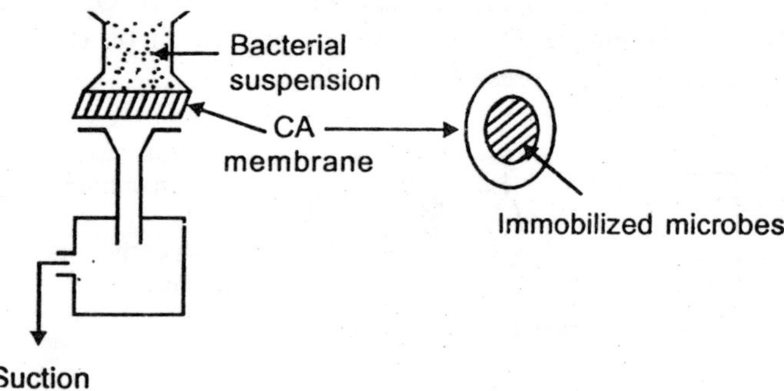

Fig. 4.8. Preparation of microbial membrane.

The advantage of sensors are that they are more tolerant to pH and temperature than the enzyme electrodes and are also comparatively cheaper and has longer life span. But the disadvantages include longer response time and cell level due to many types of enzymes selectivity in necessary. For this mutant bacterial strain deficient in a particular enzyme is more suitable.

Gas phase biosensor

The biosensors which are commercially available are used for assessing compounds in colution phase. However, recent finding of peizo-electric crystals as immuno-chemical sensors in the gas phase hold promise and convenient in area of environmental analysis. Formaldehyde can be assayed in gas phase using formaldehyde dehydrogenase layered onto a Mh_z piezo-electric crystal.

Environmental Applications of Biosensors

1. BOD_5 is widely used test in organic pollution detection. Conventional test takes 5 days of incubation. While BOD sensor using *Trichosporon*, as oxygen probe, takes only 15 minutes and life in 20 days if used continuously.
2. Microbial biosensor for detecting CH_4 and CO_2 gas been developed. For methane, immobilized bacterium *Methylomonas* is used and take 2 minutes. While for carbondioxide, *Pseudomonas* is used, and its life time is more than a month. *Thiobacillus* based sensor can detect SO_2 in environment. Co detector can help finding carbonmonoxide in mines and many industries so that fire alarm could be given.
3. Pesticide specific antibodies been used to find the presence of herbicides like triazines and carbamates in water with sensitivity of 0.13 to 0.7 mg/l within 16 minutes. For malathion the detection could be upto 16.5 ppm, for sevin 0.5 ppm and for parathion it is even upto 0.1 ppm. Immuno-electrodes are very useful when analyte concentration is low. Graphite electrode with fixed cyanobacterium *Synechococcus*, can measure electron transport inhibition in organism during photosynthesis by herbicide action. The sensitivity is found about 50 μg/l and the time taken in 10 mins.

   Use of acetylcholine esterase obtained from bovine erythrocytes may help the detection of carbamate and organophosphate group of insecticides in water. Portable pesticide monitors are available in markets of developed countries. Other biosensors against polychlorinated biphenyls chlorinated hydrocarbons and aromatic compounds be used.
4. For detection of phenols, oxidase enzyme obtained from mushrooms and potatoes have been used. Maleic dehydrogenase based fuel cell can detect methanol contamination in drinking water upto 10^{-3} ppm. This is much below the limits of regular GLC or MS analysis.
5. Heavy metals can be detected and estimated using enzymes with –SH group. Cu^{2+} is detected using the enzyme tyrosinase.

Some of biological components used in biosensor for environmental purposes are given below :

Biological component	*Physical component*	*Parameter measured*
Trichosporon cutaneum	O_2 electrode	BOD
Azotobacter vinelandi	NH_3 electrode	Nitrate
Methylomonas flagellae	O_2 electrode	CH_4
Bacillus subtilis	O_2 electrode	Mutagen screening
Mixed Nitrifying bacteria	O_2 electrode	NH_3 and NO_2
NADH and Dehydrogenase	Redox electrode	Ethanol
Antibody to Parathion	Piezo-electric crystal	Parathion (Pesticide)
Cholinesterase	Chem. F.E.T.	Nerve gas
ACH receptor	Conductimeter	Nerve gas

Thus the domain of biosensor technology is growing and basic research is concentrated on various application possibilities, selectivity, accuracy, rapidity and cost-effectivity. For detection pesticide concentration by enzyme sensor it may cost approx. $ 3.00 per test against $ 15.00 in conventional test.

Molecular film technology is being improved for a uniform planar coating of sensing material. While *hybrid sensor* concept has also developed. It has a combination of immobilized cells and enzyme membrane to have improved sensitivity. Efforts are continuously made to improve the sensing technology and field has assumed such as importance that a separate journal on *biosensor* is solely devoted to it.

In USA, different agencies engaged in environmental biotechnology development, NASA and NIST are trying to evolve better types of sensors for accurate detection of carcinogens and other dangerous toxins in soil and water. In such cases biosensors been found to be superior in sensitivity than ordinary physical instruments.

5

PREVENTION AND CONTROL OF WATER POLLUTION

Biodegradable pollutants alone cause the measure pollution load but there are also several kinds of in gradients such as heavy metals, refractory chemical compounds and mineral oil which cannot be decomposed by mechanical or biological means. Inorganic fertilizers and various biocides are too responsible for contributing to this load. Efforts of control should, be undertaken to reduce the amounts of waste by reutilizing or recycling of their components. Control of water bodies and of organisms serving the purpose of water protection should get reinforced and carried out by all available means including legal enforcement under the provisions laid down in, the Water (Prevention and Control of Pollution) Act, 1974 and the Environmental (Protection) Act, 1986. The various ways techniques suggested for prevention and control of water pollution are briefed as follows.

Stabilization of Ecosystem

It is the most reliable way to control water pollution which involves reduction in waste input, harvesting and removal of biomass, trapping of nutrients, fish management and aeration. Various physical and biological methods are followed to restore species diversification and eco-balance in the water body to prevent pollution.

Reutilization and Recycling of Waste

Various kinds of wastes like paper pulp, municipal and industrial effluents, sewage and pollutants can be recycled to advantage. For example, urban waste could be recycled to generate cheaper fuel gas and electricity. One large sine urban waste recycling plant is already existing at Okhla near New Delhi. The total gas generation from the planı is about 0.6 million cubic feet per day having a heat value of 700-800 BIU per cubic foot.

A new technology of waste recycling and disposal has been introduced by a distillery in Gujarat. This technology would not only help the distillery to treat 450,000 litres of waste daily before letting the effluent into streams but also generate energy equivalent to that given by 10 tonnes of coal every day. The regulated use of thermonuclear reactions in controlled fusion reactors to produce cheap electricity as suggested by Eastlund and Gough (1969) could solve the problem of accumulation of waste in the twenty-first century.

Removal of Pollutants

The various physicochemical techniques used for removal of chemical biological or radiobiological pollutants have been adsorption, electrodialysis, ion exchange and reverse-osmosis of the various techniques, the reverse-osmosis techniques is based on the removal of salts and other substances from water by forcing the later through a semipermeable membrane under a pressure that exceeds the osmotic pressure so that flow is in the reverse direction to the normal osmotic flow.

In practice this involves a porous membrane whose chemical nature has been such that it has a preferential attraction for solvent while repulsion for the solute. Reverseosmosis has been commonly used to desalinate brackish water and also finds suitable, effective and economical method for the purification of water polluted by sewage effluents. Techniques introduced by C.S.I.R. for the control of water pollution have been successfully used for the removal and reuse of pollutants from industrial effluents. Some of the achievements are described as follows:

Decolourization of water

An electrolyte decomposition technique has been developed to decolourise the waste water from saree dyeing and printing industries around Jetpur in Gujarat.

Removal of phenolics

Phenolics in waste water produced from industries such as pulp and paper mills, carbonization plants, petroleum refineries, training industries and resin manufacturing units are removed by the use of polymeric adsorbents.

Removal of sodium salts

Reverse osmotic technique is used to recover sodium sulphate from rayon mill effluent the technique is also used to recover water for reuse. Similarly, 70 percent of the protein and 80 percent of the lactose can be recovered from cheese whey by reverse osmosis; which otherwise would cause a serious pollution problem.

Ammonia removal from the waste water of industry

Ion-exchange technique using a weakly acidic cation exchanger based on methacrylic acid is employed to remove ammonia from waste water in the forms of ammonium sulphate which can be reduced for the manufacture of fertilizer. The process has good application potential not only for pollution abatement, but also for the recovery of the pollutant in the form of useful salt.

Removal of mercury from chlor-alkali plants effluents

Mercury thrown out from chlor-alkali plants gets removed and recovered by mercury-selective ion-exchange resin.

6

Waste Water Treatment

Most effluents (flowing) are composed of diversed elements and contain both dissolved and suspended matters be it organic or inorganic. Treatment of such waste water involves the removal of contamination and is done to prevent effects on water or allow its reuse, which is challenged by Scientists and Engineers. In the conventional treatment task, 3 steps are considered

1. *Primary Treatment*. This is physico-chemical process where sedimentation, chemical coagulation and precipitation are followed to remove the coarse and fine suspended solids.
2. *Secondary Treatment*. This involves the removal of colloidal and dissolved organic substances. It is essentially a biological treatment process where the principles of biodegradation and biosorption are used.
3. *Tertiary Treatment*. This is also called 'polishing', where purification is needed depending on the reuse of that water. At this stage, excess nutrients like nitrogen, phosphorous and toxic elements like metals are removed.

Biological treatment is mainly based on the exploitation of catabolic activities of microbes including bacteria, algae, fungi, protozoa, rotifers and nematodes. Requirements include adequate supply of microbes, ensuring their contact with the flowing, and oxygen availability. It has been adequately reviewed by Foster (1980).

Secondary treatment of waste water is done in controlled environments or *bioreactors*. It offers the possibility of shortening the treatment time and manipulation of process, although there could be natural systems of treatment in lagoons and stabilization ponds using both the microbes and higher plants. The latter takes long time and cannot be done in populated areas due to availability of adequate land.

The conventional system comprises two main processes, namely, fluidized microbial culture and film-flow system. In the former category comes the activated sludge process and the latter includes trickling filter and rotating biological contractor (RBC)

Activated Sludge Process

In an aerobic process, utilized in secondary treatment plants, where organic matter is brought in contact with diverse types of microbes in presence of a mechanical aerator. The microbes are gathered in a form of *flocs* each having a diameter of ca 0.1 mm or larger. This separates cells from treated liquid. There is a continuous implant of the mixed culture with microbes present in the incoming liquid and in feedback of cells to culture vessel. The features are aeration tank, solid/liquid separation and sludge recycle.

The flow from aeration tank containing the flocculant microbial biomass (known as *sludge*) is separated in a settling tank. In this tank, the separate sludge is without contact with organic matter. And becomes activated, and this is recycled in aeration tank as a 'Seed', while the rest sludge is a waste and disposed later. High O_2 level is maintained by constant mechanical aeration either by 'bubbling' or through the use of rotating paddles. Development of protozoa in a sludge indicates healthy condition. The sludge disposal is a problem, it is used as fertilizer in crop lands or as landfills after drying. It may also be 'Lagooned'.

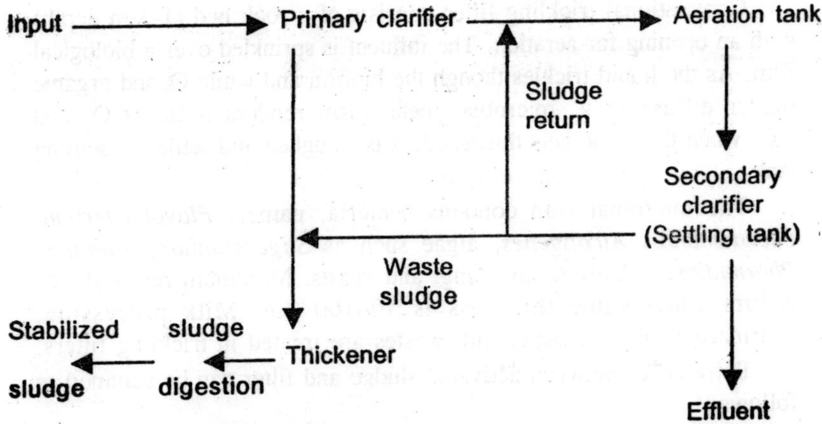

Fig. 6.1. Flow diagram of activated sludge.

The main advantage of this is its compactness for a given rate or removal of substrate. It microbes grow, a 'bulking sludge' may be formed, which hinder settling. As a result, some microbes may escape in the flowing.

Trickling Filter

This is older method and is used in small sewage tanks, which not filters, but oxidation units, where organic matter is absorbed and oxidized when it passes through the filter medium containing a microbial slime layer. The media are crushed stones or rocks, plastics and slags. First filter was built in 1890.

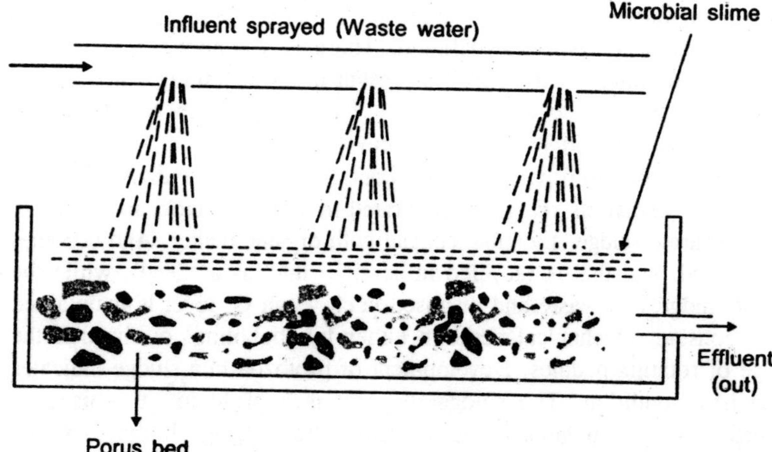

Fig. 6.2. Schematic view of trickling filter.

Conventional trickling filter consists of a rock bed (1-3 m depth) with an opening for aeration. The influent is sprinkled over a biological film. As the liquid trickles though the biofilm and while O_2 and organic matter diffuse in it, microbial metabolism renders it into CO_2 and NO_2 when the slime gets thickened, it is sloughed and settles in settling tank.

The microbial film contains bacteria, namely *Flavobacterium*, *Pseudomonas*, *Alcoligenes*, algae such as *Stigeoclanium*, *Utolhrix*, *Phormidium*, *Chlorella* and fungi and yeasts. Maximum removal rate occurs when slime thickness is 70–100 μm. Milk processing, pharmaceuticals and paper mill wastes are treated in trickling filters.

Differences between activated sludge and filter can be summed as follows :

Activated sludge	*Trickling filter*
1. Microbial growth is suspended in flocs	Biofilm is fixed on a medium
2. Solids from settlers are partially recycled	Solids are wasted
3. Effective in removing pathogens	Less effective
4. Comparatively high operating cost	Low operating cost

ROTATING BIOLOGICAL CONTRACTOR (RBC)

This is another *film flow bioreactor*. Microbial slime film is built upon a partly submerged support medium, consists of a series of light weight disks of Polyethylene/Styrofoam/PVC mounted on a horizontal shaft in a tank through which waste water is allowed to flow. The shaft is rotated slowly by a low speed motor. RBCs can be linked in parallel or in series. A rotating 'biological disk' treatment of this type was first constructed with circular steel disks and evaluated on experimental basis in 1925. Improvements in tank configurations and other parameters have been devised.

The plastic disks nowadays are of 2.3 m diameter with spacings of 20 mm between the disks which have 40% of surface area submerged. Medium is rotated at 1-7 revolutions per minute and slime biomass of 1-4 mm layer is developed. As medium rotates, the bio-film is exposed to the nutrient in waster water and air. Shearing stress cause excess biomass to be stripped off the medium in a manner

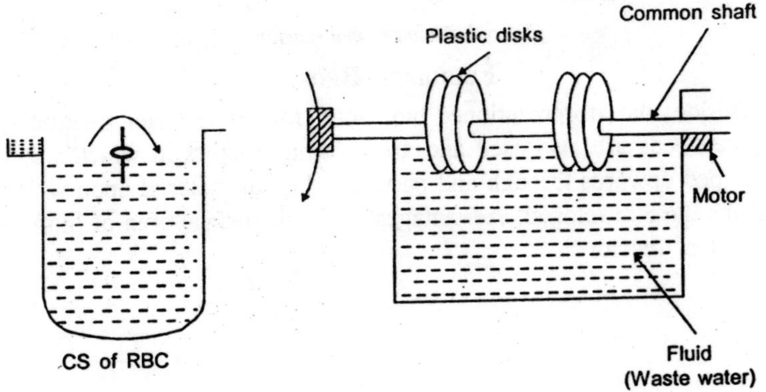

Fig. 6.3. Schematic diagram of rotating biological contactor.

similar to trickling filter. This biomass is removed in a clarifier. The design and operation for this plant has been reviewed by Antonie (1976) and the factors controlling performance by Poon et al. (1979). The treatment influenced by (1) rotational speed, (2) waste water retention time (3) staging (4) temperature and (5) sub-mergence. This type is used in treatment of municipal waste water and industrial ones like pulp, textile, meat and vegetables. It is efficient in removal of phenols from industrial effluents flows by use of suitable microbes. Its cost is less than activated sludge and is more stable, efficient and versatile. Gas-tight RBCs around for co-metabolic treatments of chlorinated solvents using methanotrophic bacteria. The disadvantages of RBC are somewhat higher BOD and higher suspended soils.

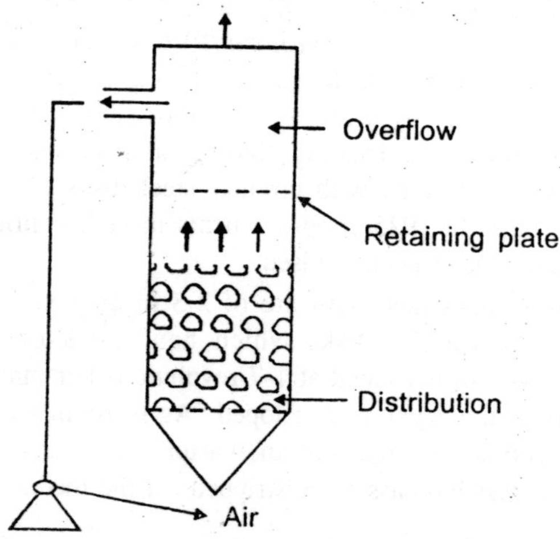

Fig. 6.4. Fluidized bed reactor.

Fluidized Beds

Besides the aforementioned bioreactor for waste water treatment, fluidized beds are also used and are column reactors in which water is pumped upwards through and particular bed of material upon which biofilms have developed. Advantages are high surface area of biofilm per unit of reactor.

7

SEWAGE TREATMENT

The most common form of water pollution control consists of a system of sewers and waste water treatment plants. Waste water is collected by the sewer system and delivered to the waste treatment plant, where it is made "fit" for discharge back into the general water supply. The discovery of more pollutants is causing serious questions to be raised concerning the real meaning of the term "fit" as used above. The methods used to treat sewage before it gets discharged into rivers, lakes, or the ocean are not excessively expensive.

The procedures are regarded to be effective for the purpose of rendering the waste moderately safe insofar as it bacterial content is concerned. But the virus loud and toxic chemical content of sewage discharges e not been as thoroughly evaluated. The processes of treating municipal sewage have been broadly classified as *primary*, *secondary*, and *tertiary*. The particular process used in a given situation has been found to depend on the volume to be treated, the location of the outfall, the dilution factor, the potential hazard to users receiving the water, and in many cases, the cost of the project.

Primary Treatment Process

Primary treatment is a mechanical process which simply removes solids, Metal screens stop large solids; sands and small stones settle in a grit chamber from which the water passes into sedimentation tank, where its rate of flow has been sharply reduced and small particles settle as a sludge. Scum at the top is removed. The quantities removed at this stage can be huge. Primary treatment is able to remove organic material responsible for 25-30 percent of the biological oxygen demand, of the sewage. However, primary treatment is normally carried out in a series of steps:

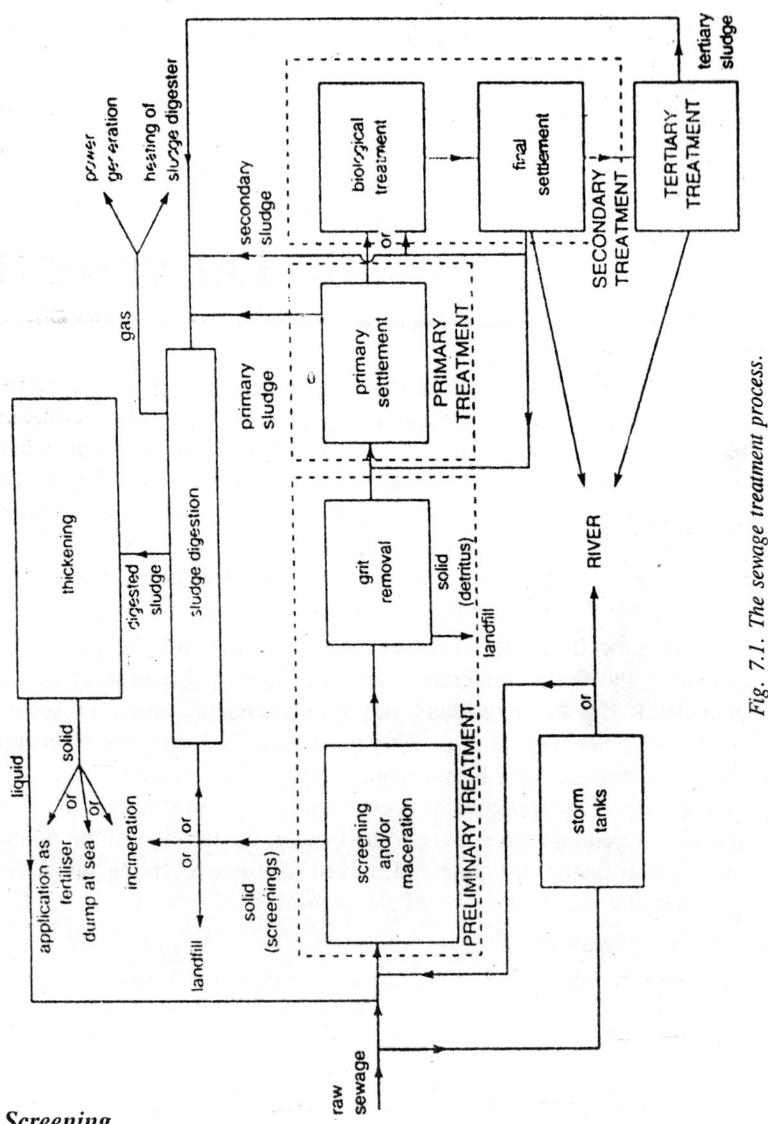

Fig. 7.1. The sewage treatment process.

Screening

Large floating objects could be removed by passing the waste water through screens. Some plants use a device called a comminutor, which screens and grinds the material. The shredded or ground material remains in the water to be removed later in a settling tank.

Grit removal

Sand, grit, cinders, and small stones are allowed to settle to the bottom of a grit chamber. It has been a very important step for cities that have combined storm and municipal sewage systems. The grit obtained in this process is disposed of by using it for land fill.

Sediment removal

Sewage, even after removal of grit, still has suspended solids. These will settle out if the speed of sewage flow gets reduced. This is complished in a sedimentation tank. The suspended solids settle out and the solid mass, called raw sludge, has been collected for disposal.

The primary treatment has been completed when the effluent, from which grit and sludge have been removed, has been treated with chlorine gas before discharge into a stream or river. Chlorine gas is added to destroy disease-causing bacteria. Primary treatment removes about one-third of the BOD and suspend solids and a few percent of the refractory (persistent) organic compounds and plant nutrients.

It is obvious that today, when concentrations of pollutants are discussed in parts per million, simple primary treatment of sewage should be supplemented by further treatment methods. The primary sludge a burdensome problem because it is; bulky and must be moved. It contains 94 to 99 percent water. Usually the first step has been to remove as much of the water as possible. In so cases, the sludge has been dried in beds with some of the water being removed by filtration. The residue gets disposed of on land. Because the sludge itself can comprise a pollution problem, a better method has been to bring about microbial decomposition in sludge digestion tank before drying Sometimes sludge sedimentation and digestion occur in a partitioned structure called an *Imholf tank*.

The microbial action lake place under anaerobic conditions, and because this proceeds slowly at low temperature the digestion tanks are usually heated to 80 to 90°F, at which, temperatures the sludge: decomposes in 20 to 30 days. The digested sludge will have been reduced to about one-third of its original volume and will be relatively inoffensive. The other products of primary treatment have been gases and the fluid or clarified waste water. The gas has been mostly methane, which is usually burned as fuel to provide heat for the digesters an other equipment The clarified waste water has highly objectionable properties and in most cases has been put through a secondary-treatment.

Secondary Treatment Process

Secondary treatment is essentially a biological process which is designed to remove most if not all remaining organic matter. Water leaving conventional secondary treatment facilities normally is down to a BOD of 10 percent or less of the initial value; 95 percent of the original bacteria is removed, along with 10 percent of the phosphates and perhaps as much as 50 percent of total nitrogen. In the activated sludge process of secondary treatment, the incoming sewage is mixed with decomposer bacteria and air or oxygen. One of the hazards of this method has been that a toxic, industrial chemical many enter the sewage, kill the decomposer bacteria and wreak the operation for weeks. The mixing of industrial wastes with town wastes at one treatment plant is thus a potential costly and dangerous practice, needing constant monitoring. Aeration (mixing with air) may be done by spraying the water over filter beds or by sending it through large, rotating, multinozzled distributor 'pipes that also let the water trickle through multilayered filter beds.

A modern filter bed has a layer of relatively large pieces, of anthracite coal on top, then a layer of smaller granules of silica sand, and finally a bottom layer of small-grain, high-density garnet sand. This operation produces a sludge that must 'be disposed of. After treatment, the water may get chlorinated to kill disease-causing bacteria. Sewage from about 40 percent of the population receives Secondary as well as primary treatment. Two processes are currently available for secondary treatment: the tricking filter and activated sludge processes. An efficiently operating activated sludge system is able to remove up to 90 cent of the suspended solids and BOD.

A good trickling filter system has been capable of removing 80.85 percent, but in practice 75 percent is more common. A trickling filter has been simply a bed of stones and gravel 3 to 10 feet deep, through which the sewage passes slowly. Bacteria gather and multiply on the stones and gravel until they become numerous enough to consume most of the organic matter in the sewage. The water, after passing through the activated bed, trickles out through pipes in the bottom of the filter. The trend in new secondary treatment plants in away from trickling filters and toward the activated sludge process. In this process, the rate of bacterial action is increased by bringing air and bacteria laden sludge into very intimate contact with the sewage which has previously received primary treatment.

Sewage, air, and activated sludge remain in contact for several hours in the aeration tank. During this time, the organic wastes are broken, down by bacterial a ion. A recent improvement in the process has been made by using pure oxygen instead of air. It has been known for a number of years that more bacteria could be supported in a smaller spa with less pumping of air (oxygen) if pure oxygen were used.

An economically competition system using oxygen has only recently been developed. The system achieves 90 percent utilization of oxygen, compared to 5-10 percent in conventional systems. It has been called by some the most significant recent advance in sewage treatment. The sewage flows from the aeration tank into another sedimentation tank, where solids could be removed. Chlorination completes the basis secondary treatment. The sludge, which has the bacteria, can be used again by returning it to the aeration tank and mixing it with new sewage and air or pure oxygen. A few of the treatment plants heat-dry the activated sludge, usually after some form or mechanical water removal, called *dewatering*, and sell the product as fertilizer. The process has been costly and there has been only a limited demand for the product. It is less expensive to incinerate the sludge in furnace or to use it as landfill.

Incineration sterilizes the sludge and reduces its volume. However, incineration is some having some disadvantages. It creates an air pollution problem and leaves an ash that must be disposed of Still, disposing of a small amount of ash is easier then getting rid of a large amount of sludge. The methods used for disposal of sludges have been usually those that are the least costly. Whether disposal on land or sea is selected depends on the proximity of the treatment plant to suitable disposal locations. Dewatered sludges are commonly used as landfill. Trucks, trains, barges, and pipelines are use as landfill. Trucks, trains, barges, and pipelines are used for transporting the sludge. Liquid sludges have been disposed of either on land or in bodies of water. Liquid sludge is used to fertilize or condition agricultural land. However, problems of odour water pollution, and stimulation of insect and algal growth, as well as other aspects of public health and aesthetic values, must be regarded.

Tertiary or Advanced Treatment Processes

Tertiary treatment is able to remove virtually all the remaining contaminants. Water leaving conventional secondary treatment still is having most of the original phosphates and nitrates, any persistent insecticides and herbicides, disease-causing bacteria and viruses (if any), and perhaps a number of industrial, organic compounds. Waste water that is not subjected to tertiary treatment contains the nutrients on which algae thrive. Primary and secondary sewage treatment lower the BOD of the water and eliminate harmful bacteria. They do not, however, effectively remove other dissolved organic and inorganic compounds.

If water is going to meet water quality standards of the federal government, (some are now in effect and others are yet to be established) attention must be paid to these modem-day pollutants. Thousands of waste treatment plants will be constructed or expanded in the future to meet the demands for pure water and these plants will be built to meet the government's water, quality standards. They will look operate much differently from the plants built during the last 30 years. During the last decade a wide variety of treatment step beyond secondary have been considered, and some are no being incorporated into water treatment sequence on a trial basis. These advanced treatment techniques, under investigation, range from extension of biological processes capable of removing nitrogen and phosphorus nutrients, to physicochemical separation techniques such as adsorption, distillation, and reverse osmosis. Most dissolved refractory organic compounds remain in water that has gone through primary and secondary treatment. These persistent compounds resist bacterial action.

The effects of such compounds in water are not all known, but taste and odour problems, tainting of edible fish, and fish kills have

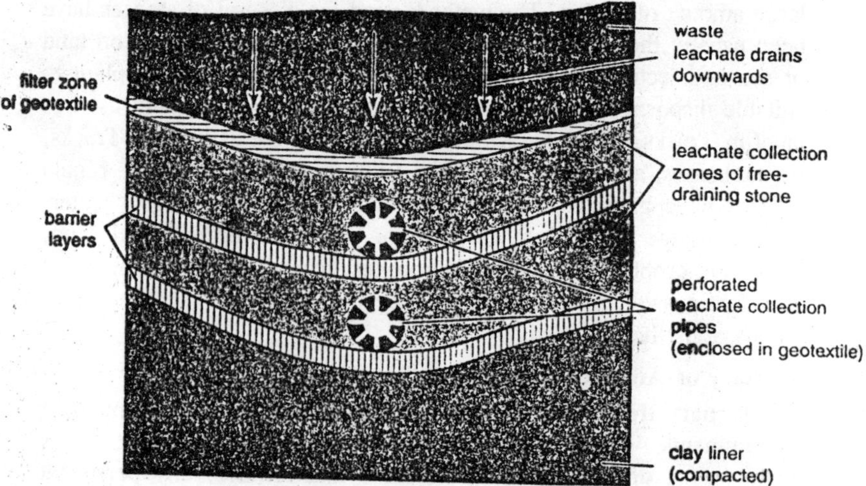

Fig. 7.2. System suitable for the lining sites.

been attributed to their presence. It has been found possible to remove 70-80 percent of these compounds by passing the previously treated water through a bad of activated carbon. The organic compounds leave the water and are absorbed on the surface of the carbon. Most of the carbon now used is in granular form but tile use of a powdered form is being investigated. The powdered form needs less contact time but is more difficult to handle. Granular carbon could be reactivated for further use by heating it in multiple hearth furnaces. A loss of about 5 percent of the carbon is experienced during each reactivation. When powdered carbon has been used, it is put directly into the water. The organics adsorb to the carbon, which is then removed by using coagulating chemicals.

The usefulness of the process depends upon the development of effective method or regenerating the used powdered carbon for reuse. The plant nutrient, phosphorus, could be removed from water by precipitation methods. This technique may be used as a separate step in waste water treatment. Two chemical approaches re commonly used: in one, lime (CaO) if added to make the water alkaline and to precipitate the phosphorus; and in the other approach, metal hydroxides are added. In either case, are inorganic phosphorus (as phosphate) is precipitate as insoluble phosphate salts of such metal cations as Fe^{3+},

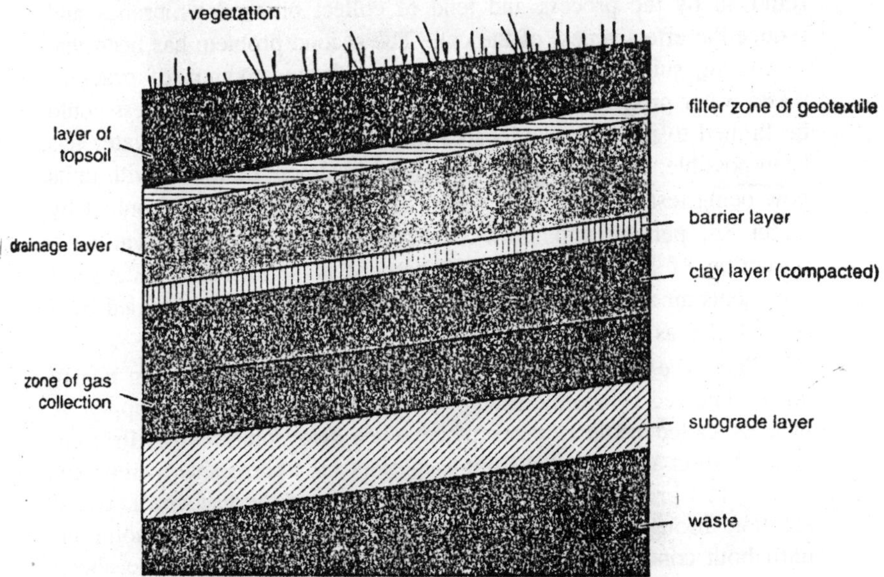

Fig. 7.3. System suitable for the capping of secured landfill sites.

Al^{+3} or Ca^{+2}, and the organic phosphorus compounds are absorbed onto the hydroxide floc (precipitate) formed by this same cations in alkaline solutions. The resulting sludge could be collected and treated to regenerate the precipitating agent. Unfortunately, nitrates, the other basic plant nutrient, cannot be removed the same was as phosphates because most nitrates have been water-soluble, Biological methods are being sought as possible solutions to this problem.

The removal of inorganic salts has been a problem which most be dealt with in water treatment. Typically a city doubles the initial concentration of Salts in water by using it. A possible method for removing these salts from used water has been electrodialysis. Electricity and membranes have been used in this complicated process. The membrane used in usually made of some chemically treated plastic. When this method is used, an electric current is allowed to pass through the water by means of two electrodes. The immersed electrodes are separated from each other by membranes.

The ions in solution are attracted toward the electrodes, pass through the membranes, and leave cleaner water behind. The treated water is ready for reuse of further treatment. The electrolysis method has two problems associated with it. Organic molecules cannot be removed by the process and tend to collect on the membranes and reduce the effectiveness of the cell. The second problem has been that of finding suitable disposal sites for the large amounts of brackish waste water produced. Because of the latter problem, the process could be limited to use in areas located near large bodies of salt water has been possible. A one-step electrodialysis treatment of water, with units now being tested, is able to reduce the total dissolved salt content by about 35, percent and allows a 92 percent recovery of water. A reduction of salt content by 35 percent is significant, because it represents an amount of dissolved salt nearly equal to that added by a typical city as a result of use.

Thus, the salts added to sewage during use in a city can nearly all be removed before the sewage is returned to the general supply. A process called osmosis takes place when two solutions of different concentration are separated from each other by a permeable membrane. During this process, water molecules flow from the less concentrated solution through the membrane, into the more concentrated solution, until both concentrations are equal. Reverse osmosis uses the above naturally-occurring process in reverse. Sufficient pressure can be exerted on the solution of higher concentration to overcome the tendency for water molecules to flow in, and instead cause them to flow out. This process amounts to removing water from the waste materials, rather than removing the waste materials from water. One of the problems encountered with this equipment has been devising a suitable support for the large surfaces of the weak membrane so it can withstand the necessary pressure.

Organic molecules tend to foul the membrane, but the problem has been not as serious with this technique as it has been in electrodialysis. As it is the water and not the ions that pass through the membrane, reverse osmosis can reduce both organic and mineral salt content of the water. Pilot plant studies reveal a 90 percent reduction of total solids and a 75 percent recovery of water. The disposal of brackish water has been a problem with this method, as it has been with electrodialysis. It must be noted that none of the processes discussed can complete the primary and secondary treatment methods. A series of treatment steps that might be applied, for example, has been as follows:

1. *Primary treatment*. Removes material that will settle or float.
2. *Secondary treatment*. Removes biologically decomposable impurities.
3. *Precipitation*. Removes phosphorus compounds and suspended solids.
4. *Adsorption*. Removes dissolved organics.
5. *Electrodialysis*. Returns dissolved salt concentration to, the level present before use.
6. *Chlorination*. Removes disease-causing organisms.

A series of steps such as these does show promise for waste water treatment in the future. Phosphates are the most difficult to remove from polluted water, yet their concentration must be reduced in areas where waste waters case lakes to undergo rapid eutrophication (e.g., the Great Lakes area). Instead of tertiary treatment, which actually disposes of nitrates and phosphates, the water from primary or secondary treatment in some it areas gets channeled onto land where these constituents will acts fertilizers and the water will irrigate crops. Another approach is to let partly treated wastewater be released into beds of sands and gravel and returned to the aquifer. A novel method of final wage treatment is worked out by the Institute of Water Research, Michigan State University.

Water from secondary treatment is piped to the first of these man-made lakes where the nitrates and phosphates in the water nourish aquatic plants. The pots, which have a nutritional value resembling that of alfalfa, may be harvested and fed to livestock. Then the water moves to second lake where catach and bluegill bass grow. The surrounding grounds are landscaped for public recreation. The water is finally sprayed over nearby land to return it to the aquifer.

Sludge Removal

The better and larger a sewage treatment plant, the more sludge it produces, and handling this material can use up half the operating budget of a treatment plant. The sludge is dried and disinfected and is to be used to make fertilizer. Most of the large cities situated on the banks of ocean send concentrated wastes and sludges into the ocean through huge pipes.

Septic Tanks

In rural areas and rapidly developed urban communities municipal sewage systems usually are absent, and home-owners install a septic tank. A septic tank is a waterproof, unventilated metal or concrete container placed underground to receive raw sewage. Bacteria act on the organic components, reduce their volume, and convert them to gases, some liquids (particular organic acids), and inert solids that settle Periodically the tank must be pumped clean. The gases diffuse into the soil, and the liquids flow into underground performed, pipes that spread out from the tank into a dispersal field. In ventilated soil, bacteria act further on organic matter. Poorly located septic tanks endanger water supplies. Bacteria need time to act. If the dispersal field is too close to a well or it local aquifer, and if the soil is too sandy, waste liquids from the tank may enter the local water supply before bacterial decomposition gets completed.

Sterilization of Water Obtained from Effluents

With the increasing need to reuse domestic effluent as drinking water methods of sterilization, are steadily assuming greater importance. Most of the effluent from sewage works has been discharged directly into rivers, and the discharge is carried out usually without sterilization. It is regarded that natural processes will be enough to destroy the bacteria still present in the sewage effluent after treatment. This assumption has been made more in hope than in certainty, and is very far from the truth.

Rivers and lakes in many parts of the world have been so heavily polluted by bacteria that they are rightly considered dangerous for bathing. The essence of the problem has been of course, that the oxygen content of most waters has been too low to have an appreciable effect in reducing the bacterial count. The principal bacteria and viruses present in purified sewage effluent released to rivers and thence to the sea have been as, follows:

1. *Dysenteric bacteria*—Flexner's *Bacillus dysenteriae*, *Shigella dysenteriae*, *Shigella Paradysenteriae* and *Proteus vulgaris*.
2. *Koch's bacillus*—which cause tuberculosis.
3. *Vibrio cholerae*—which causes epidemic cholera. (This bacillus is no longer widely distributed in temperate areas.)
4. *Typhoid fever bacilli—Salmonella typhosa*, *Salmonella' paratyphi*, Gartner's, Morgan's and Schottmuller's bacilli.
5. *Leptospira icterohaemorrhagiae*-which causes so called mud fever.
6. Polio virus.
7. Virus of infections hepatitis.
8. Enteric cytopathogenic virus-which causes epidemic diarrhoea.
9. Adenovirus-which appears to cause eye and other infections.

In addition, it becomes possible for parasitic worms of many kinds to be discharged from purified sewage, but only if filtration has been poor. Provided the filter medium has been such that all particles greater than 20 microns in diameter are removed, there should be no danger of worm infestation. Amoebae which can make a variety of conditions including liver ulcers and intestinal haemorrhage have a diameter of 50 microns, and should also be removed by efficient filtration. Bacteria, on the whole, have been very sensitive to sterilizing agents and are killed efficiently.

Viruses, on the other hand, are much more resistant to oxidizing agents. The best methods of destroying viruses has been to allow a certain time between the collection and the purification of sewage effluent; virtually all viruses are dead after about a week, because they are denatured by living microorganisms. Viruses can also be destroyed effectively by chemical flocculation, because a chemical reaction occurs between the virus protein and the metallic ions of coagulants. It is estimated that some 95-99 percent of all viruses contained in a sample of sewage effluent could be destroyed in this way. In 'purified' water, viruses can survive for a long time. It helps to explain why the spread of poliomyelitis often seems to be completely irrational. It does not affect appreciably people in underdeveloped countries who drink from heavily polluted rivers with high bacterial counts but it can affect people in highly developed countries, where all the drinking water has been carefully purified and sterilized.

The reason has been that in the polluted rivers the poliovirus is soon destroyed by microorganisms, while in developed countries careful filtration and sterilization simply eliminate the virus destroyers but have no effect on the viruses. These have been so small that they can pass through any filter medium, and are inordinately resistant to chlorine, ozone and other bactericidal agents. Modem sewage treatment processes, where crude sewage has been stored for a time prior to treatment, have been now much more effective in eliminating viruses.

Bactericidal Agents

The main methods of sterilizing water have been the following;

(a) The use of chlorine and its derivatives such as hypochlorites;

(b) Ozone dosing;

(c) Ultraviolet ray irradiation;

(d) Electrolytic methods.

Chlorines and its derivatives

For the sterilization of water either chlorine gas or one of the two salts sodium hypochlorite (NaClO) or calcium hypochlorite [$Ca(ClO)_2$] have been used. Chlorine even in small concentration, effectively kills most bacteria, mainly because it destroys the enzymes needed by the microbes to survive. Enough chlorine has been normally added so that some free chlorine remains in the water after two hours of contact. This free chlorine—even in the most minute concentration—has been readily detected by the use of potassium iodide and a special indicator. For slow sterilization a slightly acid medium has been preferred, together with intimate contact between water and chlorine for at least two hours. If the temperature of the water has been less than about 10°C an excess of chlorine compensates for a slower reaction rate.

For rapid sterilization an excess of chlorine has been used in the water and allowed to destroy bacteria and other harmful organic matter for a period of about ten minutes. At the end of that time the excess chlorine gets neutralized by the addition of either sulphur dioxide, sodium sulphite or sodium thiosulphate. As an alternative, ammonia could be added to the water, which converts excess chlorine into the chloramines—NH_2Cl, $NHCl_2$, and NCl_3. These chloramines have been quite odorless and have no specific taste, and they are bactericides in their own right. The bactericidal power has been, however, rather poorer than, that of chlorine on its own. For large plants, chlorine dosing has been effectively carried out with the gas itself.

The main drawback has been that very strict control is essential; even concentration as low as 40 parts per million of free chlorine in the atmosphere have been highly dangerous. For smaller plants hypochlorites have been much easier to handle; they can be dissolved or suspended in water prior to dosing. Viruses like those of poliomyelitis and infectious hepatitis need quite high concentrations of chlorine for destruction in excess of 0.4 mg/litre for 30 minutes or more. Koch's bacillus has been even more resistant and needs at least 1 mg/litre for on hour to ensure destruction. The remaining bacteria have been, however, easily destroyed at very low concentration. Chlorine has been not too effective in destroying larger organisms; amoebae, for example, need more than 10 mg/litre for an hour for destruction. But as amoebae are having a relatively large diameter, they could be removed easily by good filtration practice.

The use of ozone

Ozone is having the chemical formula O_3 and is made by passing air through a field of 'silent' electric discharge. Ozone cannot be stored because it disintegrates spontaneously; it must, therefore, be produced, on site from air. There have been several commercial ozonizers on the market designed to operate at about 12,000 V, obtained by means of a standard step-up transformer. The air entering the ozonizer must be perfectly dry to avoid the formation of nitric acid, which would destroy the electrodes. The ozonized air is then bubbled through water. It best to use a good power agitator to ensure good distribution. When ozone is used for water sterilization, an excess of zone must be present for about five minutes where the temperature has been in excess of 10°C and for about ten minutes for temperatures below 10°C. Excess ozone could be easily detected by the use of potassium iodide/starch papers. When used correctly, ozone has been regarded to be more effective than chlorine for sterilization. The main difficulty with ozone is that it has been only sparingly soluble in water, and therefore effective dispersion has been vital to ensure that there is good contact between the ozone and the bacterial and virus matter to be destroyed.

Sterilization by ultraviolet rays

The maximum bactericidal action of ultraviolet rays is when the wavelength of the rays has been in the vicinity of 260 nm. Ultraviolet light is commonly produced by a mercury lamp. This method of sterilization has been only commercially viable when relatively small quantities of water are to be treated. The water must be very clean, otherwise some of the ultraviolet rays have been absorbed and thus rendered inactive. In general a 25 watt U/V lamp can be used for sterilization about 2000 litres of water per hour. The water has been led past the lamp at a depth of about 150 mm.

Sterilization by electrolysis

Two techniques have been used. In the first, the anode is of silver and the cathode of either carbon or stainless steel. The electrodes must always be kept where the water is naturally turbulent. On the passage of an electric current, silver ions get released in solution, and these have a marked bactericidal effect. The quantity of silver needed for adequate sterilization has been minute-normally, of the order of 1 gram of silver to 20 m^3 of water. Faraday's laws apply to the electrolysis of silver which means that 108 g of silver are liberated by the passage of 96,500 coulombs (1 coulomb equals 1 ampere second). Hence a current of I amp flowing for 1 hour liberates $108 \times 3600/96{,}500$ g of silver=4 g of silver, or enough to sterilize about 80 m^3 of water per hour. The silver, in the form of its ion, reacts with most bacteria to form a slime which can be filtered off.

The method has been not as effective as chlorine or ozone dosing, and viruses have been much more resistant to silver sterilization than are bacteria. The second method has been involving the addition of salt (NaCl) to the water; the passage of an electric current produces free chlorine ions, which kill microorganisms in the usual way. The passage of one Faraday (96,500 coulombs) of electricity produce's 35.45 g of chlorine, and therefore an electrolytic cell which uses a current of 1 amp produces 1.35 g of chlorine per hour.

Reuse of Waste Water

Many of the molecules of water that travel down the rivers in densely populated areas pass through the human physiological system several times—possibly 10 to 40 different bodies—before reaching the ocean. Even if the drinking water has been treated to kill most of the bacteria, we are repelled by the thought of drinking in the afternoon the molecules that were flushed into the sewer by our upstream neighborus in the morning. Yet that is literally what many people who live along the waterways are doing. The rivers have been used as repositories for vast quantities of domestic and industrial waste products having an unknown fate and largely unknown biological effects. Some use is made of waste water from sewage plants for irrigation, mainly for nonfood crops and golf courses, but in some countries, untreated or poorly treated waste is used for vegetable farming.

Along part of the southern coast, pumping of well water for the burgeoning population has lowered the natural water table to the extent that underground intrusion of sea water is a serious threat to the water supply. When the underground water has been removed by pumping, the pressure is reduced and seawater fills the void by seeping into the area through underground strata of sand.

Treated sewage waste water got pumped underground experimentally to replenish the natural water supply and create an artificial barrier against seawater instruction. Treated waste water is fed into several lakes that are used for recreational purposes. Treatment involves running the secondary waste water into an oxidation pond and then into percolation beds from which the lakes are fed by underground flow. There have been problems of eutrophication. Excessive aquatic growth causes oxygen deficiencies and subsequent fish kills. Even so, the project is considered generally successful. Similar plans for using treated waste for recreational purposes have been in progress. Though no public health problems have been detected, the need remains for careful study of chemical contaminants and their potential hazard to vacationers, sportsmen, and wildlife. It cannot be assumed that tertiary treatment would render the water safe for drinking even if the effluent were free of pathogenic bacteria. Waste water may contain viruses as well as high concentrations of nitrates, phosphates, and other mineral salts.

Organic degradation products of detergents and other domestic and industrial synthetic chemicals of largely unknown composition and biological activity may be present. The only place where waste water from sewage: treatment is recycled directly into the drinking water is at Windhoek, South Africa, where treated sewage effluent normally makes up 14 percent of the water supply, increasing in 40 percent during the winter. South Africa is desperately short of water sources; most cities are unwilling to bear the expense of extensive tertiary treatment or to run the risks the undetermined hazards that remain. Methods of advanced treatment that will permit large-scale reuse of waste water for domestic use have not yet been developed. We can effectively cope with nearly all the water-borne bacterial diseases by conventional methods of treating supplies of domestic water.

Sedimentation, nitration, chlorination, or combinations of treatments are generally reliable in providing drinking water that is potable and bacteriologically safe. Pathogenic viruses, however, present a more difficult problem. The only one of the enterovirus diseases that has been proved to be water-borne has been infectious hepatitis, a dangerous liver disease that is incapacitating and sometimes fatal. The disease has been prevalent in many parts of the world. The spread of infectious hepatitis, as well as other viral diseases, by the reuse of treated waste water has been an eventuality that must be thoroughly explored before widespread reuse of processed sewage waste can be contemplated with confidence: Evidence indicates that the chlorination of secondary effluent in accordance with current practices does not remove or inactivate all of the viruses.

It has been possible to filter out some viruses by adsorption on activated carbon, but the viruses are readily released and remain infectious. Removal by adsorption on precipitates has been also under study. Irradiation by gamma rays from radioactive isotopes can kill viruses as well as bacteria, but the effectiveness of disinfactants such as high-level treatment with chlorine or other disinfectants requires further research. Many of the reaction products of such chemicals acting on the organic materials also remain to be determined. The renovation of waste water to a quality that would allow its reuse for a variety of purposes has been a major objective of current research projects.

Some industrial users can tolerate more impurities than agricultural, recreational, or domestic users. Needed improvements include removal or inactivation of viruses, removal of salts or their dilution to lower the salt content to acceptable levels, continuous monitoring for a wide variety of chemical pollutants, and probably the development of ways to remove toxic substances. Water usage is about 25 percent of the fresh water supply that is now economically available from all sources.

It is estimated that we could provide 600 billion gallons per day out of a total runoff of 1,200 billion gallons. But there would still be serious shortages in some areas, especially in the drier sections of the west. As the demand for water increases and the price goes up, increasingly costly methods for the recovery of waste water become more feasible. The problem is twofold: water in good supply and water of good quality. Future additional drinking water supplies may be obtained by several processes:

1. Greater economy in the use of water. For example, backyards and gardens are getting smaller in waterpoor areas of high population density.
2. Dual supplies, involving one source for drinking, cooking, and bathing and another source for other domestic and horticultural

uses. The latter would probably consume at least 80 percent of the domestic water supply.

3. Complete recycling-requiring, however, careful and extensive study of fail-safe systems for pollution control and removal of contaminants.
4. Desalting of brackish and ocean waters, involving, how ever, high costs and transport problems.

Sooner or later, much of the sewage water will have to be recycled in order to meet the demand.

8

Biotreatment of Wastes

Enormous range of metabolic talents exist in the microbial world. Due to natural ability many microbes can synthesize some biologically active agents like specific enzymes and biopolymers to detoxify, degrade or biotransform various toxic substances present in environment. A major challenge in todays society is to harness this biopotential in the environmental clean-up. Bioremediation and bioreclamation are new technologies and need proper application and scale up.

In recent years with growth of different ways of microbial metabolism of organic and inorganic substances and possibility of genetic control through molecular biological tools, focus of research is now shifting to genetic management to enhance or modify potential is an efficient and cost effective way. The improved microbes so produced are known as GEMs.

Genetic Concept in Environment Management

The different strains of same microbial species respond differently to environmental pollutants led to concept of presence of specific genetic determinants and the development of *genetic ecology*. This idea crystalized with observation of differential resistance capacity with toxic metals. It became clear that microbes do possess unique mechanisms to adopt themselves to various pollutants. Each organism has its own genes that help it to adopt itself in environment and use the available substrate as a source of carbon and energy. These genes could be located either in nuclear chromosomes which are linear DNA sequences and in *plasmids*, which are small extracellular, self-replicating circular DNA as in bacteria.

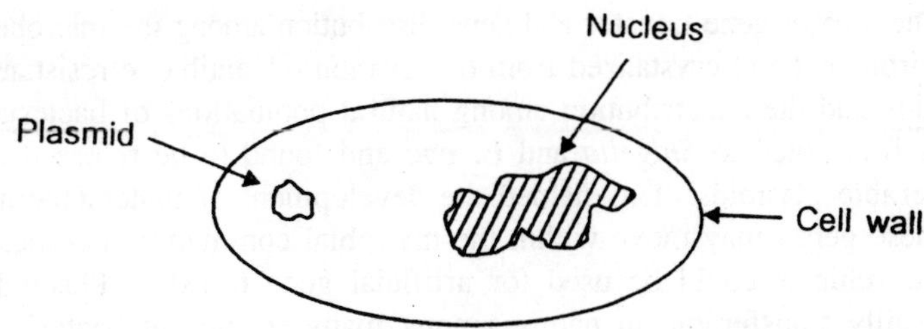

Fig. 8.1. A bacterial cell.

It is now known that in bacteria, majority of the degradative genes for organic matter are primarily located in plasmids, rather than chromosomal DNA and biodegradation of organics or dangerous wastes is essentially a plasmid mediated process as indicated in Table 8.1.

Table 8.1. Some Natural Catabolic Plasmids

Plasmid type	*Substrate*	*Organism*
CAM	Camphor	*Pseudomonas putida*
SAL	Salicylate	*Pseudomonas* sps
TOL	Tolluene, Xylene	*P. putida*
OCT	Octane	*P. olleovorans*
NAH	Napthallene	*P. putida*
pAC25	3-chlorobenzoate	
puu220	Halo-alkanates	*Alcaligenes* sp.

Micro-organisms capable of using halogen-substituted haloalkanoic acids have been readily isolated from environment. In dehalogenation process, carbon-halogen bond is cleaved by enzyme dehalogenase and the products are easily metabolized. Each of biodegrative enzyme specific for a toxic compound is controlled by a protein, which in turn is controlled by a specific gene that contains the genetic information from DNA of the degrading organism. The underlying principle is well-known 'one gene one polypeptide' concept.

In nature, it is often noticed that catabolic activity is either slow or not sufficient enough for their widespread application. This can be enhanced either by (1) selection of *constitutive mutants* (2) *gene amplification* to broaden the activities of genetic determinants (3) *gene transfer* from one microbe to the other through natural process as it often happens in a microbial community in nature or (4) by applying modern technology of molecular cloning and *recombinant DNA* (R-DNA).

Gene Transfer in Environment

The idea of gene transfer and gene distribution among the microbes in environment and crystalized from observation of 'antibiotic resistant' plasmids and their distribution among natural populations of bacteria. It was first noted in *Shigella* and *E. coli* and found to be released to transferable plasmids. This helped the development of understanding that these genes may move within the microbial community and these genetic vehicles could be used for artificial gene transfer. Plasmids are readily transferable in nature among many species of bacterium by 3 genetic processes, viz., *conjugation*, *transformation* and *transduction*. R-factor plasmids for drug and metal resistance have been found in a wide variety of microbes. Inter and intra-plasmidic genetic recombination may lead to reshuffling of genes on catabolic plasmids. Moreover, integration of plasmid-DNA with the chromosomal-DNA has been found to produce a family of new catabolic genes in a *Pseudomonas syringae* strain. Genetic homology of mercury resistance among *Citrobacter* and *Klebsiella* strains have also been noted.

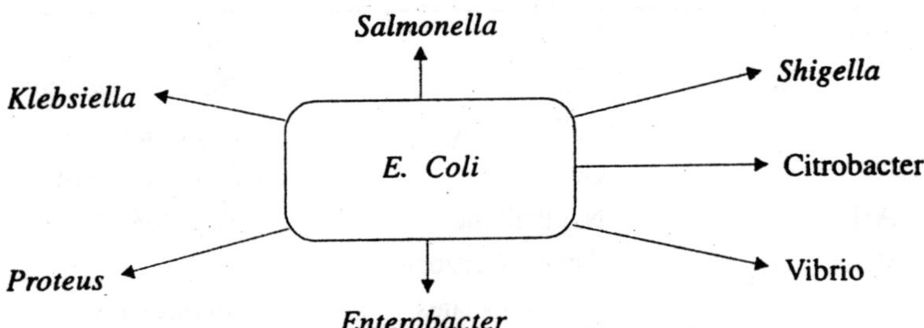

Fig. 8.2. ENT plasmid transferred and maintained in a variety of genera.

ENT plasmids (*Enterotoxigenic*) found in *E. coli* have also been found to be present in a variety of bacteria genera of same group. They have been found to be transferable and expressable in many gram negative bacteria.

Among metabolic plasmids, transposable nature of TOL plasmid has been noted, while genetic relatedness among 2, 4-D and 3-Cba plasmids has been observed in *Pseudomonas*, indicating horizontal gene transfer.

Genetically Engineered Microbes (GEMs) in Biotreatment of Wastes

Accepting possibility of gene transfer among the microbial communities, attempts to improve the degradative capacity *in vitro* by applying modern day sophisticated technology of genetic engineering have been made. A.M.Chakraborty's pioneering work in use of *Pseudomonas putida* as a biodedgrader of petroleum hydrocarbons in oil spills is a forerunner of many exciting later discoveries of use of GEMs. Hydrocarbon degrading genes occur in clusters in portion of possibility of genetic transfer in other non-biodegradative microbes has opened up a new vista of biotreatment of wastes.

During last few decades a host of new chemicals, which are new to environment and normal microflora, have been injected. These are not only toxic to biota, but refractory as well as to normal biotransformation. This led to belief that genetic restructuring of normal biodegrading microbes to design new catabolic pathways is deemed desirable.

It is known that through process of *gene technology*, segments of degradative genes of plastids can be cut and transferred to another bacterium lacking the same. This depends on harnessing the natural genetic exchange mechanisms to engineer the required organism. The plasticity of catabolic plasmid has provided the 'raw material' of genetic exchange for creation of novel organism with higher degradative potential to tackle the novel substrate. The recombinant DNA (R-DNA) can multiply in host bacterium and confer specific degradative capacity.

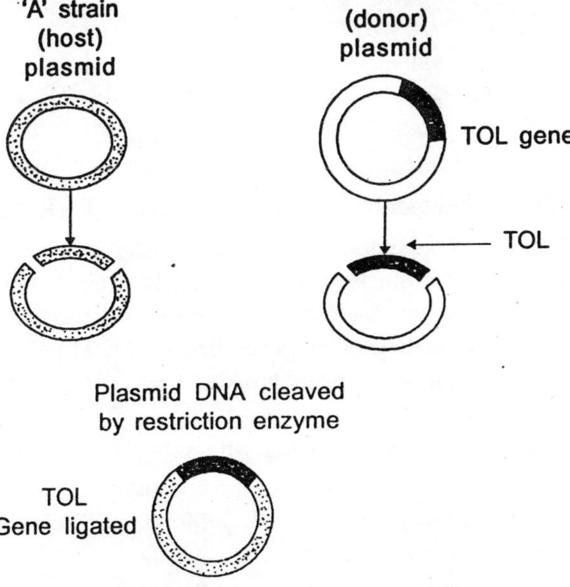

Fig. 8.3. R-DNA with TOL locus in 'A' plasmid.

Alternatively, in the same strain the specific gene can be amplified to augment its degradative power.

GEMs have now become a tool to biotechnologists. They are now not only useful in industrial production of various substances of biological origin and in agricultural practices to improve crop production through gene cloning, but also to exploit vast potential of degradation / detoxification of various dangerous substances, which were earlier through to be not easily bioamenable, through the creation of a host of 'super bugs'. Chakraborty and his group has constructed a multiplasmid *Pseudomonas bacterium* and patented it for first time. It was ***first living organism*** which was allowed to be patented by the American Supreme Court. Through presence of a number of oxido-reductive or hydrooxylation enzymes, the bacterium can now degrade a number of hydrocarbon molecules and / or aromatic compounds which are highly toxic to environment.

2,4-D is biodegradable. But 2,4,5-T is recalcitrant, perhaps due to additional halo-substitution. Kellog, 1981 by their 'plasmid assisted molecular breeding, could isolate *Pseudomonas cepacia* strain AC 1100 which harboured *two* plasmids and could finally degrade 2,4,5-T.

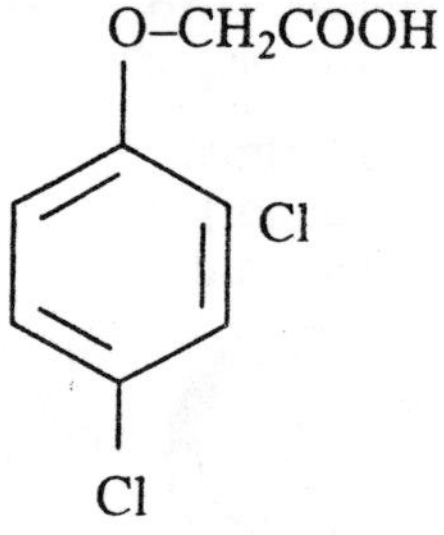

$O\text{–}CH_2COOH$ (with Cl at positions 2 and 4)

2,4-D (Biodegradable)

$O\text{–}CH_2COOH$ (with Cl at positions 2, 4 and 5)

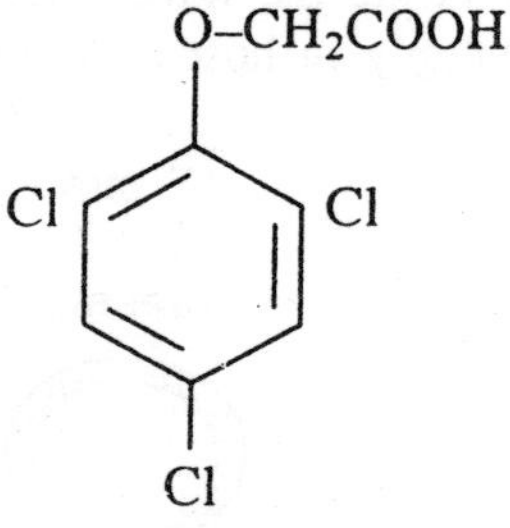

2,4,5-T (Recalcitrant)

It has been possible to genetically transfer TOL plasmid and PCB degrading genes from *Pseudomonas putida* strain LB 400 to *E. coli*. Multiplasmid system has also been constructed through hybridization with plastids encoding other genes like CAM, HAA, OCT, etc., so that the resultant strain with broad range degradative power could grow in crude oil.

Some other potential GEMs with substrates are listed below :

GEMs	Substrate
Pseudomonas putida	Mono and Dichloroaromatics
P. oleovorans	Alkane
P. diminuta	Parathion

P. cepacia	2,4,5-T
Alcoligenes sp.	2, 4-D
Acinetobacter sp.	4-chlorobenzene

Vannillate from paper industry and SDS from detergents can be degraded by genetic 'tinkering' of *Pseudomonas* sp. Even lignin degrading plasmid genes have been identified in bacteria from wood and paper pulp effluents.

$$\text{SDS} \xrightarrow{\text{SDS-gene in Pseudomonas}} \text{Dodecanol} \rightarrow \text{Acetate}$$

Most of these genes are available in microbes of waste sites and need to be isolated for the development of efficient biodegraders through process of genetic engineering/*gene cloning*. It has been noticed that in this respect, the gram negative bacterial genus *Pseudomonas* is a versatile one in providing a lot of degradative genes which can be cloned to others.

These GEMs have variety of used in reclamation of toxic wastes sites including land fill areas, battle fields, mine effluents for metal removal and oil-spills from drilling sites and accidental leaking from oil tank. In many cases, the pollutant may not serve as a direct source of good to GEM or its surface change may not attract the microbe or it may be hydrophobic and unavailable to microbe. In the later case, the GEM has to synthesize a molecule that would emulsify or solubilize the toxicant for easy availability. In that respect a strain of *Pseudomonas aeruginosa* was constructed by A. Chakraborty and his group, which could produce a glycolipid emulsifier to reduce the surface tension of an oil water interface and help the degradation of oil. Bio-surfactants so produced are ecologically safe also.

Genetically engineered microbes may also be structured in such way that they either synthesize extracellular polymers as in *Zoogloea* to bind with toxic metals such as Cr, Cu, Co, Ni, etc., and help precipitate them from polluted water or synthesize phosphatase enzyme on cell surface to cause metal precipitation from solution, as in *Citrobacter*.

So GEMs may prove to be 'biotechnological wonders' in coming decades including environmental management, even though their biosafety has been questioned.

Fate of GEMs in Environment and Bio-Safety

The engineered GEMs, so constructed have to be subjected to screening for eco-sustenance i.e., their ecological stability, persistance, proliferation and display of improved performance. This is because they have to function in a foreign environment which is often competitive with indigenous microflora and hostile environmental factors may not be conducive to survival and character expression. This poses a challenge to microbial ecologists. They are required to be monitored regularly with modern molecular genetic probe techniques.

There is a possibility of hazard risks in environment with use of GEMs. This is a controversial and often bitterly debated issue. Whether the release of such novel organism is advantageous or dis-advantageous would depend on its features and those of receiving ecosystem. There is a general fear that new organism may *alter* the pattern of microflora in local environment and thus disturb the ecological balance. In certain cases they may turn into *rogue pests* and affect the habitat. The basis of this lies infact that while altering the degradative features, some undesirable changes may occur within the bacterial cell and it may turn virulent affecting even mankind (a *genetic bomb*).

Some of common microbes which are subjected to genetic alteration for environmental clean-up and for which we need to have more ecological information are *Pseudomonas*, *Alcaligenes*, *Vibro*, *Acinetobacter*, *Klebsiella*, *Enterobacter* and *Nitrosomonas*.

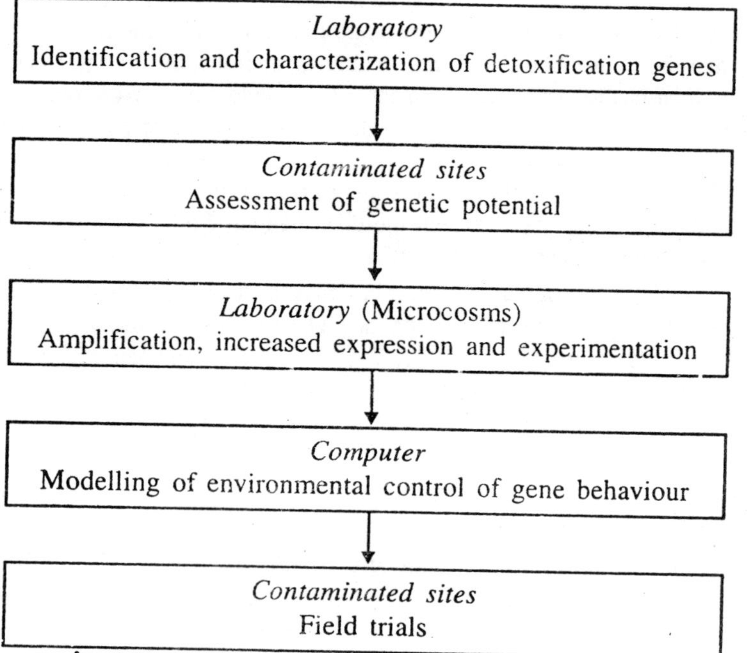

Fig. 8.4. Strategy for handling GEMs in detoxification of wastes.

To avoid risks in release of GEMs: The following steps may be taken :

1. Gain maximum information about donor and recipient organisms.
2. Genetic exchange schedule to highly specific.
3. Recipient organism should be locally adapted.
4. Proper identification of 'marker' character should be there in the altered organism to trace it in environment.
5. Constant monitoring of its fate and interactions in the new ecological niches is absolutely necessary.

Potential risks can be minimized by working in a controlled environment and to that effect, large bio-reactors can provide a means of assuring a safe biodegradation. Genetic ecological research for detoxification of wastes may be summed up. Therefore the practice is to first test them in *microcosm* experiment in laboratory to analyze the dependencies of genetic potential and expression efficiency. Then in *mesocosm* for field trial before final release. It is also possible that such GEMs still be modified to disarm them ecologically so that they may spread beyond a defined region.

9

AIR POLLUTION

The atmosphere forms an insulating blanket around the earth. Without it the temperature at the equator would rise to 180 during the day and drop as low as—220°F at night. It burns up meteors that would bombard the surface of the earth from space. Without the atmosphere there would be no sound and no flight. There would be no conventional long-distance radio communication, for this is dependent on the electrons in the upper atmosphere. Without air there would be no lightning, no clouds, no wind, no rain, no snow, and no fire. The surface of the

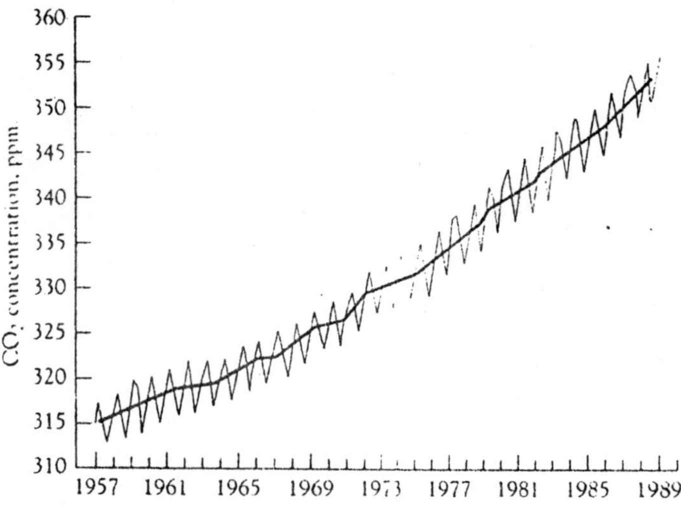

Fig. 9.1. Monthly mean concentrations of atmospheric CO_2 at Mauna Loa, Hawaii, 1958-1989. Similar trends are evident in Alaska, American Samoa, and the South Pole.

earth would be as bleak and sterile as the moon. The atmosphere shields of the earth from lethal concentrations of ultraviolet radiation. It selectively filters the sun's rays so that only two small segments of the electromagnetic spectrum penetrate to the earth's surface in appreciable amounts.

One segment is the *optical window*, consisting essentially of the visible spectrum of light, from near-ultraviolet to near-infrared. The other is the *radio window*, consisting of radio waves from about 1 centimeter to 40 meters in length. It is impossible to define the limits of the atmosphere because the atmosphere becomes progressively tenuous with increasing distance from the earth. There is no boundary between the atmosphere and the void of outer space. However, 75 percent of the earth's atmosphere lies within 10 miles or the surface and 99 percent of the atmosphere lies below an altitude of 30 kilometers (19 mites). The total mass has been estimated at 5,500 trillion tons.

The Zones of the Atmosphere

Based somewhat on the way in which the temperature of the atmosphere changes with increasing altitude, scientists have defined zones of the atmosphere. Each zone is more or less a spherical shell of the atmosphere. The thickness of a zone varies with latitudes and with the season of the year.

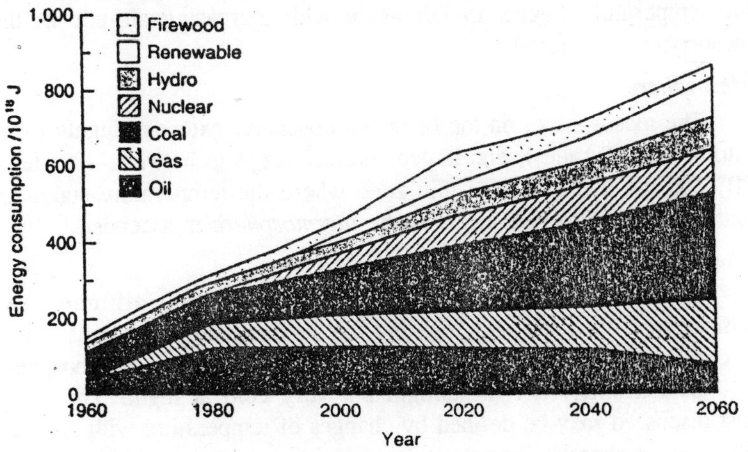

Fig. 9.2. Projected worldwide trends in source of fuel.

Troposphere

The zone nearest the earth, the troposphere, is named from the *Greek tropein* (to turn, to rotate, to change). One property that changes in the troposphere is the temperature. It drops to—55°C (—65°F) by the time an altitude of about 16-18 km (10-11 miles) is reached. The rate of temperature drop is about 6.4°C per 1000 meters (3.5°F per 1000 feet), a figure called the normal environmental lapse rate. The region through which this lapse rate occurs defines the troposphere. At the equator the troposphere is about 18 kilometers (11 miles) thick, but at the poles it is only about 8 kilometers (5 miles) thick. Essentially all the atmosphere's water vapour is in the troposphere which therefore encloses essentially the storms and precipitation of the earth. The jet stream occurs at an altitude of 10.5-15 kilometers (35,000-40,000 feet; 6.5-7.5 miles).

Stratosphere

This zone starts at the top of the troposphere, at a thin atmospheric "shell" called the tropopause, where the temperature begins to stay fairly constant, increasing slightly, in going to an altitude of 50 kilometers (31 miles). The stratosphere includes much of the *ozone layer*. The warming trend in the stratosphere is caused by a cycle of chemical changes in the ozone layer which converts all the incoming high energy ultraviolet rays to heat (see Ozone). If these rays reached the earth's surface life as we know it would be impossible. The *stratopause* marks the narrow zone at the top of the stratosphere where the temperature begins to fall again with increasing altitude as the mesosphere is entered.

Mesosphere

The mesosphere, on top of the stratosphere, extends roughly to 80 kilometers (50 miles), and its temperature drops to between—80 and—90°C at the mesopause, a thin zone where the temperature stabilizes and soon starts to rise again as the *thermosphere* is ascended.

Thermosphere

This zone is above 80 kilometers (50 mites) altitude. The temperature becomes very high but the air has such a low density that it can hold very little heat. Any, denser object in the thermosphere will be extremely hot in sunlight but very cold at night. The zones just discussed may be defined by changes of temperature with altitude. Changes in chemical composition or activity, also occur with altitude. Up to about 80 kilometers (50 miles) the proportions of components given in Table 9.1. remain nearly constant, and the long zone up to that altitude is sometimes called the *homosphere*.

Within it is a region called the *chemosphere* extending from the upper troposphere where, a huge number of chemical changes occur powered by solar radiation. Ozone production is one such activity. Above the homosphere lies the *heterosphere*. For its first 40 kilometers (25 miles), the ionosphere, the very thin concentration of atoms and molecules exists largely as electrically charged particles called ions. This globe-enveloping band of ions reflects outgoing radio waves back to earth, making radio transmission possible well beyond the horizon. Television waves are not reflected, and a TV transmitter's range does not extent beyond the horizon Meteorites entering the earth's atmosphere generally out in the ionosphere.

Table 9.1. The Gaseous Composition of Unpolluted Air (Dry Basis)

	ppm (vol)	*$\mu G/m^3$*
Nitrogen	780,900	8.95×10^8
Oxygen	209,400	2.74×10^8
Water	—	—
Argon	9,300	1.52×10^7
Carbon dioxide	315	5.67×10^5
Neon	18	1.49×10^4
Helium	5.2	8.50×10^2
Methane	1.0-1.2	$6.56\text{-}7.87 \times 10^2$
Krypton	1.0	3.43×10^3
Nitrous oxide	0.5	9.00×10^2
Hydrogen	0.5	4.13×10^1
Xenon	0.08	4.29×10^2
Organic vapours	ca. 0.02	—

The stratosphere has been of interest to aeronautical scientists because it is traversed by airplanes; to communication scientists because of radio and television communications; and to air pollution scientists because global transport of pollution, particularly the debris of above ground atomic bomb tests and of volcanic eruptions, occurs in the stratosphere and because absorption and scattering of solar energy also occurs in the, stratosphere. The troposphere has been the region in which we live and is the region to which this book is primarily devoted.

Atmospheric Pressure (Barometric Pressure)

A column of air whose base is one square inch and which extends outward beyond the thermosphere weighs at sea level 14.7 pounds on the average, when the surface air temperature is 0°C.

STANDARD ATMOSPHERE

1. An arbitrarily agreed-to vertical distribution of atmospheric pressure, temperature, and density used in altimeter design and, ballistic calculations 2. A pressure unit with a value of 1013.2 millibars, 29.9213 inches of mercury (750 millimeters of mercury) the standard atmospheric pressure at 0°C and under the standard value of the gravitational constant, 980.665 centimeters per second.

UNPOLLUTED AIR

The gaseous composition of unpolluted tropospheric air is given in Table 9.2. Unpolluted air is a concept, i.e., what the composition of the air would be if man and his works were not on earth. We will never know the precise composition of unpolluted air because by the time we had the means and the desire to determine its composition, man had been polluting the air for thousands of years. Now even at the most remote locations, at sea, at the poles, in the deserts, and mountains, the air may be best described as dilute polluted air. It closely approximates unpolluted air, but differs from it to the extent that it contains vestiges of diffused and aged man-made pollution.

Table 9.2. The Gaseous Composition of Unpolluted Air (Wet Basis)

	ppm (vol)	*μG/m³*
Nitrogen	756,500	8.67×10^8
Oxygen	202,900	2.65×10^8
Water	31,200	2.30×10^7
Argon	9,000	1.47×10^7
Carbon dioxide	305	5.49×10^5
Neon	17.4	1.44×10^4
Helium	5.0	8.25×10^2
Methane	0.97-1.16	$6.35\text{-}7.63 \times 10^2$
Krypton	0.97	3.32×10^3
Nitrous oxide	0.49	8.73×10^3
Hydrogen	0.49	4.00×10^1
Xenon	0.08	4.17×10^2
Organic vapours	ca. 0.02	—

The real atmosphere has been more than a dry mixture of permanent gases. It has other constituents-vapour of both water and organic liquids; and particulate matter held in suspension, Above their temperature of condensation, the vapour molecules act just as permanent gas molecules in the air. The predominant vapour in the air has been water vapour. Below its condensation temperature, if the air is saturated, it changes from vapour to liquid. We are all familiar with this phenomenon since it appears as fog -or mist in the air, and as condensed liquid water on windows and other cold surfaces exposed, to the air. The quantity of water vapour in the air gets varied greatly from almost infinite number of possible organic compounds that may be in the air have been identified. The major constituents of air, nitrogen (78%), oxygen (20.94%) and argon (0.93%), do not react with one another under normal circumstances.

Similarly, the trace components helium neon, krypton, xenon, hydrogen and nitrous oxide have little or no interaction with other molecules. A number of other gases also present in trace quantities have been not inert chemically but interact with the biosphere, the hydrosphere and each other and so have a limited residence time in the atmosphere and characteristically variable concentrations. It is this reaction group of gases which are considered pollutant when they are produced by, man in sufficient quantities for the background concentrations of Table 9.3 to be significantly exceeded. The most important gases in this group have been those which are universally present in the air of the world's cities, namely, sulphur dioxide (SO_2), nitrogen oxides (NO and NO_2, carbon monoxide (CO), and non-methane hydrocarbons. Other reactive gases can also bring about pollution problems at elevated concentrations, for example, the halogen gases chlorine and fluorine and their acid derivatives hydrochloric and hydrofluoric acid, but these problems tend to be local rather than universal, and their background concentrations an order of magnitude or more less than the gases reported in Table 9.3.

POLLUTED AIR

Man's activities take place, for the most part, on the earth's surface within the first 2 km of the atmosphere. The pollutants produced by these activities get injected directly into the troposphere where they are mixed and transported. The background concentrations of the reactive gases have remained, to the best or our knowledge, constant with time.

It implies that the sources and sinks (as the formation and removal processes are generally called) are in balance, and also,. for gases with a high pollutant contribution, that the sinks are able to cope with the additional burden from man. It could be partly explained by reactivity and partly by the fact that these gases natural source greatly exceeds

Table 9.3. The Composition of Dry Air in the Lower Troposphere (Free of Water Vapour)

	Chemical Symbol	*Concentration (a) %*	*Calculated Residence Time*
Principal Gases			
Nitrogen	N_2	73.0	Continuous
Oxygen	O_2	20.9	Continuous
Argon	A	0.93	Continuous
Carbon Dioxide	CO_2	0.032(b)	20 years (c)
Trace Gases			
(a) Permanent Gases (non-reactive)		Ppm	
Helium	He	5.2	Continuous
Neon	Ne	18.0	Continuous
Krypton	Kr	1.1	Continuous
Xenon	Xe	0.086	Continuous
Hydrogen	H_2	0.5	?
Nitrous Oxide	N_2O	0.25	8-10 years
(b) *Reactive Gases*			
Carbon Monoxide	CO	0.1	0.2-0.3 year
Methane	CH_4	1.4	< 2 years
Non-Methane Hydrocarbons	HC	0.02	?
Nitric Oxide	NO	0.2 to 2.0×10^{-3}	2.8 days
Nitrogen Dioxide	NO_2	0.5 to 4.0×10^{-3}	
Ammonia	NH_3	6 to 20×10^{-3}	1-4 days
Sulphur Dioxide	SO_2	0.03 to 1.2×10^{-2}	1-6 days
Ozone	O_2	0 to 0.05	?

Notes:

(a) This is the atmospheric background concentration, and not the concentratịons found in polluted areas. When a range of concentrations is given, it indicates that these have been measured by different workers at different places.

(b) Minimum concentration of CO_2 measured away from centres of population. In population centres, CO_2 concentrations vary from about 0.034 to 0.035 percent.

(c) For photosynthesis. Turnover time with the deep ocean is in the order or centuries

? Indicates that little is known about the residence time of the gas.

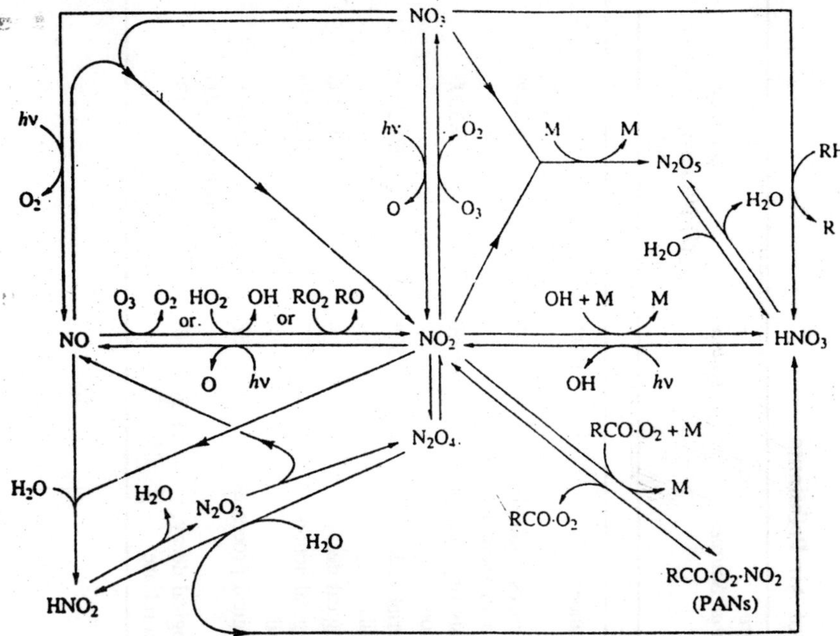

Fig. 9.3. The principal tropospheric chemical transformations of NO_x.

the pollutant one in the majority of cases (Table 9.4), It can be seen from Table 9.4 that nature produces over 10 times as much hydrogen sulphide (H_2S), at least an equivalent amount of nitrogen oxides (NO and NO_3 written as NO), and over 100 times as much ammonia (NH_3) as is produced by man. Sulphur dioxide (SO_2) looks like an exception to the rule, however. Hydrogen sulphide (H_2S) has been ultimately converted to sulphur dioxide (SO_3) in the atmosphere and is, therefore, a source of SO_3. When it is taken into account, together with the different relative molecular masses of H_2S and SO_2, it can be demonstrated that the natural and man-made sources have been equivalent. Pollution problems associated with the gasses of Table 9.4 (CO_2 excepted) arise not because of the magnitude of the man made (anthropogenic) emission but because this emission gets concentrated in the areas where people live and work, and more specifically in the cities of the industrial world.

Further, most of the world's industry has been concentrated in the northern hemisphere (over 90%), and the great majority of this between the latitude of 30°N and 60°N. In this region, the anthropogenic emission has been considerably more important than the natural one, and the

Table 9.4. Source of Air Pollutants

Gas	*Source* *Major Pollutant Sources (Anthropogenic Sources)*	*Natural Sources*	*Pollution*	*Quantity (X 10⁴ toms per annum) Natural*
1	2	3	4	
Sulphur Dioxide (SO_2)	Combustion of coal and oil, roasting of sulphide ores	Volcanoes	146	6-12
Hydrogen Sulphide (H_2S)	Chemical processes, sewage treatement	Volcanoes, biological action in swamps	3	30-100
Carbon Monoxide (CO)	Combustion, principally motor car exhausts	Forests fires terpene reaction	300	>3000
Nitrogen Oxides (NO_2)	Combustion	Bacterial action in soils	50*	60-280*
Ammonia (NH_2)	Waste treatment	Biological decay	4	100-200
Nitrous Oxide (N_2O)	Indirectly from use of nitrogenous fertilizers	Biological action in soil	>17	100-450
Hydrocarbons	Combustion, exhausts, chemical processes	Biological processes	88	CH_4 : 300-1600 Terpenses : 200
Carbon Dioxide (CO_2)	Combustion	Biological decay, ocean release	1.5×10^4	15×10^4

*Expressed as tons NO_2

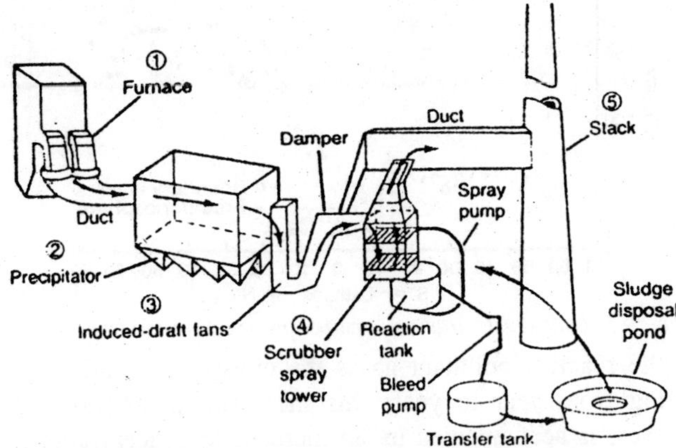

Fig. 9.4. Inside a coal scrubber. Coal is burned in the furnace or boiler (1). Fans (3) pull resultant gases through the precipitator (2) where fly ash is removed. The damper directs gases to the scrubber spray tower (4) where a slurry of water and chemical is sprayed to remove SO_2 and remaining ash. Clean gases then go up stack (5). Liquid chemical used to absorb SO_2 drains into the reaction tank where sulphur is removed through a chemical process. The bleed pump routes it to the transfer tank from which is drains into the sludge disposal pond.

potential effects have been more serious because of the concentration of the receptor of most vital interest to us—man himself. Natural emissions have been only rarely concentrated in limited areas in this way and when they are, as in the case of volcanoes, their intermittent nature and scattered location have been found to minimize the effects of their reactive gas emissions. With the exception of volcanoes, natural emissions do not get fluctuate significantly from year to year, although they may fluctuate considerably within a year. Man's emissions, on the other hand, have been steadily increasing as populations and industry expand.

The estimated worldwide production of the past century of one such pollutant-sulphur dioxide was about 5 million tonnes, mainly from coal combustion. Today, about 190 million tonnes get produced, of which about 100 million is from coal (which generally contains about 0.5 to 4 percent sulphur), 50 million tonnes from the refining and burning of oil, and the remainder chiefly from the smelting of copper, lead and zinc ores. Similar curves could be drawn for all other anthropogenic gases, despite the increased use of pollution, control measures. In the long term our most serious pollution problems have been not arising from the reactive gases but from unreactive emissions, like carbon dioxide which are not having harmful interaction, with living systems.

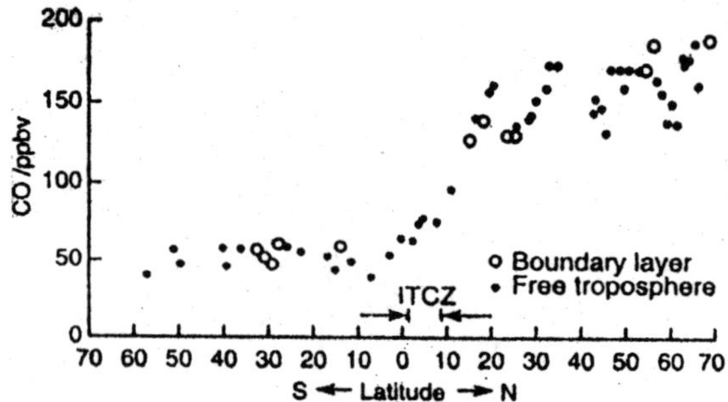

Fig. 9.5. Latitudinal variation in carbon monoxide concentrations.

Like the reactive pollutant gases the emission of carbon dioxide is increasing from year to year; this increase, in contrast with the reactive gases, is accompanied by an increase in concentration on the pollutant. Carbon dioxide plays a vital role in the earth's radiation balance and, therefore, in the earth's climate. The change in the CO_2 concentration, it is predicted, will be able to produce an associated change in the climate of the earth. Any inert pollutant will tend to increase in concentration with time if there have been no sinks available. However, this is rarely the case in practice. For CO_2 the problem has been the lack of balance between the sources and sinks over a time scale that is relevant to the atmospheric concentration (Table 9.3).

Achievement of equilibrium with the oceans would need centuries if all anthropogenic emission ceased. For that other group, of inert pollutants, and chlorofluoromethanes, whose concentration is also increasing, it has been the nature of the final removal process in the stratosphere which is the cause of concern. Although the chlorofluoromethanes appear to have no tropospheric sink, they get decomposed in the stratosphere, giving products which react with ozone. Because this removal process exists, the concentration of chlorofluoromethanes at a steady state of usage would eventually reach and equilibrium concentration in the troposphere. However, it would take decades rather than years. The real atmosphere is also having particulates and vapours, both of which are having natural cycles in the atmosphere which get affected by human activities. Man produces particles directly because of his agricultural and industrial activity, and indirectly because of atmospheric reactions of anthropogenic gas emissions.

Although the total particulates from natural sources, these particles get concentrated (like the pollutant gases which are their major source) in the industrial regions of high population density. Particles are having an effect on amenity by producing soiling, and on health by acting as carriers into the lung of trace substances like lead or the polyaromatic hydrocarbons. In the atmosphere, particles act as reaction centres, facilitating gas reactions and also adversely affecting visibility and sunlight penetration. The major vapour phase atmospheric species has been water. This persists as a vapour until supersaturation occurs, when it condenses, the vapour concentration for condensation being a function of the temperature. In practice, the water content of air has been very variable and ranges from 0 to about 4 percent by weight.

Many industrial activities emit water vapour directly—thermal power generation for example—and others effect the water cycle indirectly by release of heat to the atmosphere or to large bodies of water. Other substances also exist in the vapour phase in the atmosphere, in particular the non-methane hydrocarbons from both pollutant and natural sources, and sulphuric acid, the product of sulphur dioxide oxidation. Both vapours get associated with the incidence of smog in 'favourable' circumstances. For gases and particles to be air pollutants it is known that they must exceed their normal background concentrations to a significant extent.

It can be put in another way, substances in the air are called pollutant when their concentrations are sufficient to have adverse effects

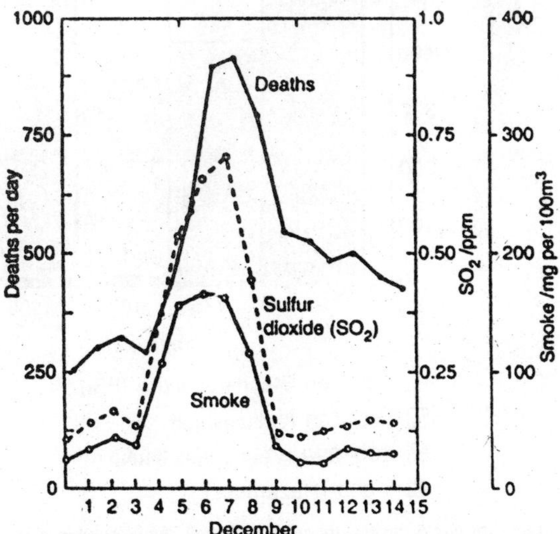

Fig. 9.6. Death rates and atmospheric pollution.

on man and his environment. Indirect effects arise because of changes in the physical properties of the earth's atmosphere system—the infra red radiation balance in the case of CO_2, and the amount of incoming ultra violet radiation in the case of ozone depletion caused by chlorofluoromethanes. The potential effects have been global in their implication and as such require international rather than national cooperation for their study and, if necessary, for their solution. Direct effects take place because of the interaction between a pollutant and a receptor such effects results from the higher than, background concentrations pertaining near the source of pollutants.

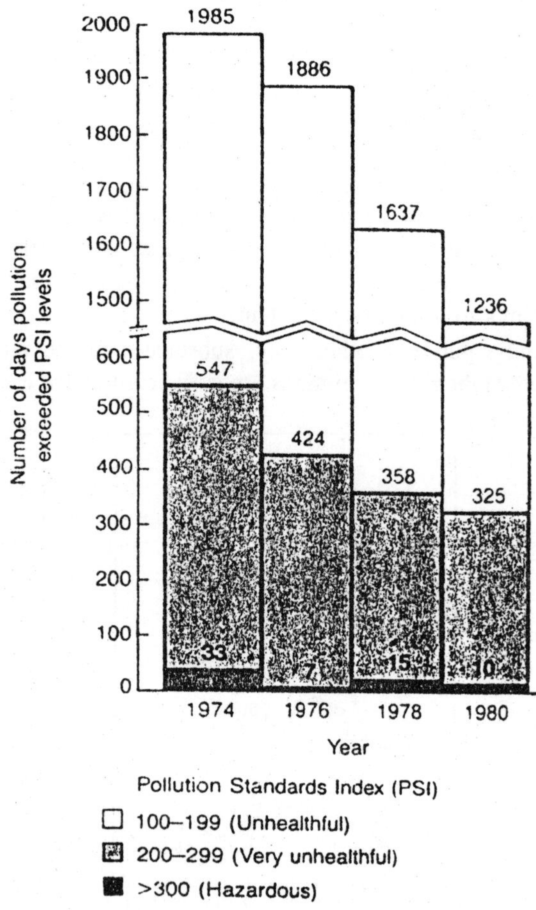

Fig. 9.7. National trends in urban air quality. Graph shows combined total of number of days that air quality was poor in twenty-three metropolitan areas.

Direct effects can be immediate if the concentration has been high enough (acute effects), or can develop over the long term as a result of continuous exposure to higher than background levels (chronic effects). The extent of area affected distant from the source or sources has been a function of the distance the pollutant is transported before the concentration is reduced, by reaction or removal at the earth's surface, to a level at which no direct effects take place. To date, it is the direct form of air pollution of which we have been most aware, particularly in its most acute manifestation—the London smogs of the 1950s. It was direct pollutian, by hydrochloric acid gas from the alkali industry, which led to the first Alkali Act in the UK in 1863 and most control legislation since that date.

Where the pollutant emission has been typical of the process, as in the case of the alkali industry, and there has been only one plant carrying out the process in the area, then an association between effects and pollutant is readily made and control measures can be taken where necessary. Other examples of unique pollutants from particular processes would be hydrogen fluoride (HF) from fertilizer plants and the aluminium industry, and odourous reduced sulphides (hydrogen sulphide and mercaptans) from the wood pulping industry. The products of fuel combustion like particulates, sulphur dioxide, nitrogen oxides, carbon monoxide and hydrocarbons are common to most industries and, in addition, are having domestic and vehicular sources, making it more difficult to lay blame for observed effects individually.

In cities, pollution comes from a multiplicity of sources, not least of which is the motor car, which contributes more than 50 percent of the hydrocarbons and nitrogen oxides to most urban airsheds and 90 percent of the carbon monoxide. These pollutants influence the city producing them most seriously under low wind speeds and stable atmospheric conditions, which prevent the dispersal of pollutants and maximize their opportunities for reaction. Reactants and products have been forming a complex mixture which can have an additive effect on a receptor or result in a significant enhancement of effect over and above what would be expected from the sum of the individual pollutants present.

10

Marine Pollution

The International Oceanographic Commission (IOC) for the United Nations Educational and Scientific Commission (UNESCO) defines marine pollution as: "Introduction by man, directly or indirectly; of substances into the marine environment (including estuaries) resulting in such deleterious effects as harm to living resources, hazards to human health, or hindrance to marine activities and reduction of amenities." If we allow broad interpretation of the word "substances" this definition appears to be all-encompassing. Note however the strong contrast between the dictionary and UNESCO definitions. The connotation of the oceanographers is that unless man introduces the pollutant, there is no pollution. Apparently the inference is that if pollution is man-caused it can be man-cured. Nearly 71 percent of the earth's surface is covered by oceans, which together comprise a total of approximately 1.37 x 10^{39} litres, weighing about 1.40 $\times$ 10^{36} metric or long tons.

This is the tremendous amount of water and consequently for many years the oceans are regarded an ideal place to dump all of man's wastes. For example, the rivers of the world discharge close to 2.0 x 10^{14} metric tons of water into the ocean every year but this is much smaller than the water already into the ocean. Carried along with the water each year are 4.5 $\times$ 10^{11} metric tons of dissolved materials. Although this is a large amount of material compared to that already present it is literally a drop in the bucket. Unfortunately, the ocean does not receive all of its dissolved material input from the world's rivers. Some comes from the atmosphere as it is washed by falling rain. Most of the pollutants placed into the atmosphere and up in the oceans either directly on indirectly from precipitation. This also is a sizeable amount of material. It is to be noted that this is larger than the amount of sediment plus the amount of dissolved materials carried to the oceans by rivers. The more important reason for being concerned about marine pollution stems from the fact that materials put into the oceans by man's activities is not evenly spread throughout the oceanic volume, pollution is generated where man lives and works and therefore finds its way for the most part into the oceanic areas that are closest to their activities.

Consequently, the dilution factor is very small and the concentrations of these pollutants build up in coastal regions. They are increased to a much higher level than would be the case if the material were distributed throughout the entire oceanic volume. Coastal are dumping grounds areas having a much higher pollution concentration not only because the material is being put into these relatively shallow areas much more rapidly than it is being carried away by natural eve nations but also because the normal structure of the oceans tends to prevent the mixing of these inputs with the rest of the oceanic volume. The oceans are not dead from pollution, not are they insensitive to the effects of pollutants. As long as we find it necessary to use the sea for a world dump, and there is no reason to believe we will ever slop, we must learn as much as we can about oceanic processes so that the assimilative capacity of the oceans is never exceeded.

Strictly speaking marine pollution may also be of natural, origin. For example, it may be caused by underwater gas and oil eruptions, or by the intensive growth and subsequent death of certain microorganisms, which are responsible for the familiar disastrous 'red tides'. However, such processes and effects are not regarded as 'pollution', even though they are often studied by methods similar or those used for investigating pollution effects caused by man. Ecological studies on marine pollution fall into three principal categories:

1. Biogeochemistry of pollutants.
2. Marine ecotoxicology.
3. Biological principles of anti-pollution measures.

The biogeochemistry of toxicants includes the investigation of the sources of pollutants, the pathways along which they enter the marine environment, patterns of accumulation in the biotic and abiotic components of ecosystems, mechanisms and rates of migration of pollutants, their transformations and other processes which determine the fate of toxicants in the sea. An important, though not decisive stage in such investigations is the determination of the content of toxicants in marine organisms and communities; this is a difficult analytical problem because the chemical composition of matter is complex, while the concentration levels of pollutants are low.

The methods of chemical oceanography are used in order to obtain and analyse samples, the results being interpreted in a biogeochemical sense. The second type of research work carried out on this problem, which is best described as 'marine ecotoxicology', involves the stud of biological effects and consequences of human interference with the composition of the marine environment. Finding answer to such questions is not by itself tantamount to solving the problem of pollution, but nevertheless yield an objective evaluation of present-day ecological anomalies in the World Ocean and predication of those likely to occur in the future. These investigations provide a scientific foundation on which to base national or international effort to preserve the seas from pollution. Solutions of the immediate practical problems concerning pollution of the sea must be based on the results of closely inter linked ecotoxicological and biogeochemical studies.

Marine Pollutants

In Table 10.1 a list of common pollutants associated with the marine environment have been given. These are not listed in any particular order of importance because it is found that in certain areas certain pollutants are more worrisome than other while the opposite condition may be true in other areas. This difference in importance comes from local conditions like population pressure, flushing, and climate.

Table 10.1. Pollutants Associated with the Marine Environment

Pathogens	Toxic organics
Sediments	Petroleum
Solid wastes	Nutrients
Heat	Radioactive materials
Fresh water	Oxygen demand materials
Brine	Acids and bases
Toxic inorganics	Aesthetical displeasing materials

Theoretical calculations of the concentrations of the principal toxicants in epipelagic waters of the World Ocean indicate that considerable pollution (lead pollution, in particular) of surface waters by atmospheric pollutants may have, taken place. This is true of inland seas. The global pollution zone of the marine environment may be characterised as follows: highest pollution levels in the euphotic layer; high pollution levels in the neritic zone and inland seas; latitudinal distribution of toxic pollutants; mosaic distribution pattern of toxicant concentrations in water; localisation of toxicants in the hyponeuston arid benthos biotope and a coincidence of the maximum pollution zones with zones of high biomass and high productivity.

Before we discuss the marine pollutants one by one we shall give some general conclusions about these:

1. The general picture of the effects and relative toxicities of the most common toxicants on natural communities of marine phytoplankton does not significantly differ from that obtained for cultures or single species.
2. The form of microcomponents in sea water are largely determined by the supersaturation of surface waters by calcium carbonate; since the suspended and the colloidal forms of this compound are strong sorbents.
3. The rate of input of petroleum and chloro-organic toxicants into the World Ocean is faster than their rate of decomposition under natural conditions. The degradation processes are particularly slow at great depths and the latter must be considered as a depot of toxic substances and their metabolites.
4. The rate of biological accumulation and migration of micro-amounts of metals and radionuclides in the World Ocean is directly correlated with their physicochemical properties and with the specific surface area of aquatic organisms.
5. The concentration levels of artificial radionuclides (^{90}Sr and ^{137}Cs), heavy metals (mercury, lead and cadmium) and chloro-organic compounds (DDT, polychlorinated biphenyl, etc.), in commercially important marine products vary with, the degree of pollution of marine biotopes, taxonomic identity of the organisms, and their ecological and physiological characteristics (trophic level, in the community, diet, age, size, etc.).

   The most variable concentrations in industrial species are typical of substances with cumulative properties, DDT and DOT metabolites, polychlorinated biphenyls and other chloro-organic compounds, and methyl mercury.
6. Typical accumulation factors of ^{90}Sr and ^{137}Cs in commercial species are of the order of 10^1—10^2 for heavy metals 10^2—10^3, and for chloro-organic compounds 10^3—10^4. The food chain effect—i.e. the dependence of the toxicant concentration in a given

organism on the trophic level of the organism in the community—is definitely present only in the cases of mercury, polychlorinated biphenyls, DDT and other chloro-organic compounds. The contents of mercury and of chloro-organic compounds in fish show a regular increase with the size (weight, age) of the animal.

7. The selective, irreversible accumulation of high concentrations of mercury in pelagic predatory fish and marine mammals is a natural process, and is not usually due to human factors.
8. Concentrations of metals in sea water and in plankton tend to decrease with increasing atomic number of the metal.
9. Global concentration levels of radioactive and chemical toxicants in commercial marine products are usually much lower than the national and international norms for maximum permissible concentrations, and thus present no danger to human health. However, in view of the possibility of local pollution and human interference with the natural background of microcomponents in individual regions, concentrations of toxicants in commercial species and marine products must continue to be constantly monitored, especially for heavy metals and chloro-organic compounds.
10. The relative content of artificial radionuclides in oceanic seston does not exceed 1 percent of their total content in water; the corresponding figure for heavy metals and transition is 1-10 percent, while in inland seas it may be as high as 25 percent.
11. The contemporary radio-ecological status of the World Ocean is distinguished by the relatively stable contents of ^{90}Sr and ^{137}Cs in surface waters and in aquatic organisms. Another typical feature is the gradual increase of radionuclide contents in the ecosystem of inland seas.
12. Self-purification of surface waters by biosedimentation of impurities is mostly due to their accumulation in aquatic organisms at low trophic levels. If the accumulation factor in seston is 1000-10000, the rate of vertical biogeochemical migration of a given impurity is of more or less the same order of magnitude as the rate of its hydrological transport (for biomasses and productivities considered as average for the World Ocean). This is true of many radionuclides (^{144}Ce, ^{95}Zr, ^{106}Ru, ^{65}Zn, ^{54}Mn, etc.), of heavy metals and transition metals present in micro amounts (mercury, lead, cadmium, iron, zinc, etc.) and chloro-organic compounds (DDT, PCB, aldrin, lindane, etc.).

13. Low concentrations of toxicants (0.1—10 μg/l) acting for a short period of time tend to stimulate photosynthesis and the rate of cell division in algae. This may be caused by the overall activation of vital processes by small doses of poisons or by an alteration of the metabolic interrelationships with the cells of the microflora. In the case of petroleum products such stimulation may be due to the presence of biologically active substances in petroleum.
14. A typical feature of the geochemical behaviour of petroleum and chlorinated hydrocarbons in the World Ocean is that their distributions in biotic and abiotic components of the same ecosystem are altogether different.
15. The inhibitory effect of the pollutants increases in the following sequence: oil, detergents, metals (arsenic, lead, cadmium, copper, mercury, methyl mercury), chloro-organic compounds. The most highly toxic compounds are DDT, polychlorinated biphenyls, mercury and methyl mercury, which inhibit photosynthesis in long-term experiments at concentrations as low as 1—10 μg/l. Arsenic, when present as arsenate ion in concentrations of up to 1—10 mg/l. had no effect on the rate of division of algae cells.
16. In a number of cases the intensity of photosynthesis of natural phytoplankton communities decreases in the presence of lower toxicant concentrations than in experiments carried out on monocultures. The higher sensitivity of natural phytoplankton to contaminants is probably due to interference with the interspecies relationships in the community.
17. There is a generally valid relationship (a direct correlation in the case of radionuclides) between the concentration levels of toxicants in commercial animals and the pollution level of the marine environment as shown by the increasing sequence of pollutant concentration: pelagic ocean waters, neritic zone, inland seas, fresh-water basins.
18. The varying responses of different monocultured species to toxic impurities in short-term experiments (less than one day) are probably due to the differing rates of absorption and penetration of toxicants into the cells. In long-term experiments the specific differences are usually much smaller and depend on molecular, cellular and tissue level mechanisms. A general characteristic is the better resistance of freshwater protococcal algae to most pollutants, as compared with cultured marine phytoplankton forms.

19. Of the species studied, *Gyrodinium fissum* is highly sensitive to mercury, the response being irreversible; the same applies to diatomous algae acted upon by a mixture of DDT and petroleum products.
20. The distribution of taxi cants in the organs and tissues of commercial fish species to typically non-homogeneous. ^{90}Strontium, lead and cadmium become preferentially concentrated in ligaments and bones; chloro-organic substances and methyl mercury concentrate in the liver and in the lipid fractions; ^{137}caesium concentrates in muscles. Chloro-organic compounds concentrate in fatty tissues.
21. The inhibition of photosynthesis is a more sensitive reaction of monocellular algae to toxicity than is the decrease in the rate of cell division.
22. Generally speaking, monocellular algae my react to the presence of toxic components in the aqueous medium in two different ways:
    (a) By inhibition of vital processes and production.
    (b) By stimulation of cell development and cell division, which is usually followed by impaired vital activities.

    The development of micro-algae is relatively independent of the presence of the toxicant in the medium.
23. Toxic (algicidal) and threshold concentration ranges are determined for various types and species of monocellular algae. Out of the several combinations of the effects of several factors studied, synergistic effects were observed for the combinations of petroleum with DDT, and combinations of certain metals with a detergent. A combined effect of several metals may be additive, or else may be manifested by a dominant inhibitory effect of the most toxic one (mercury, copper).

The considerable geographic, seasonal and other fluctuations in the reaction of natural phytoplankton to the presence of toxicants are caused not only by the environmental factors, but also by the special features of its specific composition. This composition undergoes significant changes if the action of the contaminants on the phytoplankton is sufficiently prolonged. *Nitzschia* diatoms in marine phytocoenoses are more resistant than other species to certain heavy and transition metals.

Pathogens

The pathogenic materials are the living organisms that can produce sickness or biological unbalance in either plants or animals within the ocean itself or in humans who either come into contact with oceanic waters or eat the organisms caught in the water. These include a wide variety of bacteria, protozoa, viruses and fungi. The most common of these are normally found in sewage. However, other pathogens may occur in non waste disposal areas where environmental conditions are such that the proper conditions exist for growth and reproduction.

Sediments

Sediments are always present to some extent in the marine environment. These sediments have a marked effect on plant growth because they block out a large portion of the light normally reaching greater depths and therefore decrease photosynthetic activity. In some cases rooted plants get completely destroyed by this process. As sediments are deposited on the bottom they will cover up bottom dwelling organisms (benthos) such as oysters and may even smother them under extreme conditions. The deposited sediments tend to make the water shallower so that, in regions traversed by ships, periodic dredging is needed.

Solid Waste

The disposal of solid waste has been a critical urban problem because areas suitable for the dumping of these voluminous materials are becoming scarcer. Due to this reason the ocean has been used as a dumping ground for solid waste. The solid waste most frequently marine dumped is sludge material left over as a by-product from domestic sewage treatment. However the unused products of industry and the used-up products of society are also put into the sea. If the product discharged at sea contains materials which may be leached into the oceanic environment, a serious problem could arise especially when the materials are toxic. However, when the solid materials are inert a little forethought may result in some benefit to the marine environment. Artificial fishing reefs, for example, have been found to be very successful in certain areas, serving to enhance the habitat area, especially for small fish.

Excess Heat

Excess heat, if added to the marine environment, alters, ambient conditions, and these changes may be detrimental to the organisms present. The amount of heat that is deterimental and the extent of the degradation is determined by a number of factors. The primary source of this heat has been of course, from electrical generating plants, whether they be fossil fueled or nuclear powered.

Freshwater and Brine

Although fresh water may be in great demand ashore, too much of it in the ocean obviously will produce a marked environmental, change within a small area. This change may occur due to poorly designed storm drainage systems, water diversion networks associated with dams, and effluent of some industrial processes. Excessive fresh water usually is not a critical problem. The introduction of brine into the marine environment is similar in its effect to that of fresh water except in the opposite direction. The organisms acclimated to a particular salinity now find themselves in a more saline environment which could make permanent damage. Brine is a by-product of disalination, plants.

Toxics

Toxic inorganics have been materials commonly used in industry, many of them, are relatively harmless; however, in larger quantities they may be quite destructive. There are perhaps, 35 to 40 commonly used, toxic inorganics and these should be very closely controlled by the user. The toxic organics have been the most disturbing of the modern day chemicals commonly discharged either purposefully or accidently into the marine environment. These include the biocides such as fungicides herbicides insecticides rodenticides and also the additional organics including halogenated hydrocarbons, petroleum and industrial chemicals.

The most disturbing toxic organics have been the pesticides such as, DDT and kepone which have the unfortunate characteristic of' being more soluble in oil than in water so that they tend to collect within the fatty tissues of marine organisms. They have been also very stable compounds which do not deteriorate very easily over a long period of time. These find their way into the ocean both as manufacturing effluent and as runoff after utilisation. Their long term effects are unknown upto this fine.

Petroleum

Although petroleum may be classed as a toxic organic, it is, a naturally occurring material and is biodegradable, given enough time. Its effects are not too well known. Petroleum enter into the marine environment due to accidents such as tanker damage or transfer loss, natural seepage, off shore production losses, losses associated with refineries, from runoff originating as drippings or disposal of used automobile lubricants, and unburned hydrocarbons emitted into the atmosphere, as internal combustion exhaust.

Nutrients

Nutrients (commonly called fertilizers) are those chemicals which are required by plants. The activities of man have added to the total nutrient load of almost all coastal areas. When nutrient levels becomes out of hand, plants grow unchecked so that decaying plants exist in such great numbers that the oxygen supply becomes rapidly depleted. These nutrients are present in domestic sewage effluents, agricultural runoff, and the little understood but apparently important non-point source runoff" from urban areas.

Radioactivity

Radioactive material have been not only discharged to the marine environment by nuclear power plants, nuclear power plant fuel production and reprocessing plants, and uranium activities of all sorts, but also result from more common activities such as the burning of coal. When coal is burned it gives out radioactive particles to the atmosphere which are the washed into the sea at a greater rate than any known nuclear power plant at this time. Other sources of radioactivity include the natural background, weapons testing, mine drainage accidental spillage and a few isolated industries. One of the partially unsolved problems associated with the use of nuclear energy for electrical power has been the long-term storage or disposal of spent fuels. This storage must be in an area such that there have been no pathways back to man and the deep oceans have been suggested as meeting this criterion.

Oxygen Demand

Oxygen demand materials have been those which need oxygen for degeneration and therefore steal oxygen which would normally be utilized by marine animals. Thus if too many oxygen demand materials are kept in the marine environment, the animal population will be markedly decreased due to the lack of oxygen. Sewage sludge and any other organic waste material, even that resulting from excessive plant growth due to an over supply or nutrients have been examples of common oxygen demand materials.

Acids and Bases

The discharge of acid and bases to the marine environment is quite disturbing to the natural ecological balance of the system. The normal pH of oceanic water has been somewhere around 8.0, slightly basic. This is maintained by the carbonate system. If a large amount of acid or base is introduced into the system, the carbonate reactions will be offset and an important element of the environment will get affected. In addition, there are large synergistic effects associated with pH. Most toxic materials, for example, increase their toxicity under conditions of low pH. The sources of acidic or basic materials have been primarily industrial with some of this material reaching the marine environment from accidental discharges and the repturing of tankers.

Aesthetic Considerations

Aesthetically displeasing materials include all the stuff one finds in the ocean which is unpleasant to look at or to smell. Tar balls, floatables, gas (often hydrogen sulphide) producing materials, and colouring agents are some examples of pollutants offending the senses. Although in some cases, these materials do not pose any real threat to the ecology of an area, when the area is being used for recreation, the quality of the surrounding becomes some what important.

Management Problems

Although the major pollutants are delineated, the management of marine pollution is quite complex. The reason for this is that the marine environment is used by man in many different ways and this multi-use is having a number of ramifications. In the first place each oceanic activity pursued by man may produce a whole range of pollutants. A moving ship, for example, may not only discharge pollutants in the form of waste products from the people living abroad, oil from pumping its bilges, and heat from the discharge of its condensers, it may also need deepening of a harbor channel. This may generate a salinity change in addition to sedimentation problems associated with the maintenance dredging required. Each pollutant may come from many different users.

The nutrients which primarily appear to come from the effluent of sewage disposal plants may come from storm drainage of urban areas in the form of waste products of domestic animals, along with runoff from agricultural areas that have been fertilized by either commercial fertilizers or farm animals. An example of the effect of one pollutant on another has been the attraction of heavy metals to suspended sediments. Most of the heavy metals in coastal waters are swept out of the water by sinking sedimentary particles and end up in the bottom deposits. If the heavy metals then enter their way into bottom dwelling organisms which, in turn, find their way into a food chain pathway to man, then this obviously is not desirable. Another example of this synergistic effect of two or more pollutants has been the effect of oil on Pesticides, combined with sediments.

The pesticides are more soluble in oil than they are in water so that if there is oil percent in oceanic water, it will tend to concentrate any pesticides that also may be present. Oil is also having an affinity for sediments and as it collects on the suspended particles, their weight gets increased to the point where they sink to the bottom. Portions of the inshore marine environment thus are covered with sediments very rich in both oil and pesticides. The introduction of foreign material into the marine environment has been always detrimental to man or to the environment either. In many areas the increased erosion produced by construction activities in large areas of siltation where these eroded sediments have been deposited in the shoreline areas. After stabilization, populations of marsh grass have been seen to spring up in many of these newly created areas. In the main there is no planning for this type of rehabilitation, but recent experimental work reveals that production of marshlands may very well be a common tool of the estuarine planner of the future.

Sources of Marine Pollutants

Now that the pollutants are delineated and described very briefly, we will concern ourselves with the activities involved in the production of these pollutants in an attempt to pinpoint the source of the problem. These activities are outlined in Table 10.2.

Table 10.2. Pollution Producing Activities

Marine Commerce
Industry
Electrical power generation
Sewage treatment
Other non-industrial wastes
Recreation
Construction

Marine Commerce

Marine commerce has been deeply involved with ships, and ships require a deep enough water environment so that they can pass to and from on their appointed rounds. This usually requires continuous dredging. Dredging is an expensive proposition requiring large expenditures of money. Once the sediments are extracted from the bottom and the channel is cut to the desired depth, the problem becomes one of disposing of the dredge spoils, the material dredged from the bottom. Various solutions have been suggested buť the problem is still one that has not been adequately solved. Another aspect of the ship problem has been oil. Whether these ships has been transporting oil or using it for fuel and lubrication, there always seems to be some lost of the sea. Most harbors are having large storage areas open to the atmosphere so that a certain amount of leaching into the marine environment is inevitable. Lastly, there has been a whole family of pollutants associated with any ship. A ship is just a little city afloat and any problems found in urban areas will be found on ships also.

Industry

Whenever one mentions pollution to the average individual the first source of pollution that pops into mind has been industry. Most industries require the use of water. This water has either heat or some chemical added to it before discharged into the environment. These chemicals may be anything from toxic substances to harmless colouring agents. Electrical power generation is responsible for waste heat and certainly this is one of the by-product of producing electrical power whether it be from the burning of fossil or nuclear fuels. An additional problem involved with heat exchanges has been the fact that many of these are made from copper, a toxic material which is used in anti-fouling paints on boats for many years.

Sometimes these heat exchangers are allowed to remain in an unused condition for extended periods of time with water sitting in the heat exchanger tubes. Some copper leaches into this water from the tubes and when the plant gets turned on again this copper-rich water is pumped into the environment where it may bring about some concentration among the local marine organisms. Large amounts of water pumped at high speeds from a coastal regions are able to trap or entrain small or sometimes even large organisms. These creatures may suffer mechanical damage as they go through the system, or they may get exposed to large amounts of heat for a short period of time.

Radioactive wastes are considered as one possible disadvantage of nuclear power plants, but the amount of radioactive waste actually transferred by cooling water to the surrounding environment has been measured as extremely low. The spent fuel problem appears to be of more concern at this time. The pollution problems associated with mineral extraction could not be documented due to the small opportunity for data taking, but it is expected they will be similar to marine construction. At this time the major oceanic mining effort has been directed toward the extraction of sand and gravel in the near shore areas.

Within the imminent future it is expected that a concerted effort will be made to extract manages nodules from the ocean floor, perhaps at depths as great as 3000 metres. This mining effort will be expected to have some effect on the environment but just how much has been not quite clear. The same can be said for the increase in activity connected with offshore oil drilling although it appears that there is less petroleum lost into the environment when the oil is obtained from an offshore well and transported to the continent by pipeline than when it is obtained from a land well and transported by means of tanker to its eventual user point. The spillage resulting from the use of tankers has been greater than that resulting from the use of pipe lines.

Sewage Treatment System

The process treating human waste product has been quite successful in removing suspended materials and pathogens, but it has been extremely difficult to extract all of the nitrogen and phosphorous compounds. Consequently, sewage treatment facility effluent is having reasonably high amounts of these nutrient which push plants toward uncontrolled growth. In addition, when the facility gets overloaded, as during a heavy rain, it is often completely bypassed, discharging raw sewage directly into the water. This needs a great deal of oxygen for decomposition so that the biological oxygen demand is increased. One aspect of sewage treatment neglected up to now, primarily because the costs have been so overwhelming, has been the treatment of storm runoff.

Unfortunately the solution is not as straightforward as simply increasing the capacity of existing plants to handle the heavy loads associated with storms, or even to add treatment to storm drainage systems when they exist separately. Much of the urban and rural runoff does not go through conventional pipe systems, so additional system would have to be built if all runoff has to be treated. Storm drainage washes urban streets clean. All material deposited on the streets, including crankcase-drippings, rubber worn from auto tires, and bird, dog, and cat droppings are washed to sea.

Modern studies have indicated that this runoff has been a very significant source of coliforms and petroleum, to name just two pollutants of some concern to man. How to handle these non-point sources has been a task presently facing marine environment managers and one that appears to have no inexpensive or simple resolution. With most large urban areas disposal of the solid residue from sewage treatment plants has been a solid waste disposal problem of the first magnitude. Ocean dumping in the near-shore coastal regime is utilized for many years and, to a certain extent, will probably continue to be. Either the sludge must get treated more thoroughly before it is dumped or new deep sea dumping areas must be found. If new treatment or disposal techniques are not followed, there may occur major changes in the coastal environment.

Agricultural Sources

The largest single class of non-industrial polluters are farms. Agricultural pollutants have been commercial fertilizers, animal wastes, pesticides, herbicides, and sediments. In many cases the agricultural contribution of all these pollutants has been greater than any other single source. The pollution can be stopped, and there have been cases in the past where particularly bad problems have been solved.

Recreational Activities

Usually when we think of marine recreation we think of a non-polluted area since recreation is essentially limited by pollution. In shallow water motor boats stir up the sediment on the bottom, a condition which often persists for long periods of time. This increase in turbidity markedly influence the natural plant growth and probably has something to do with the decrease in the numbers of rooted plants reported in the last few years. The aspect of motor boating that has many investigators most concerned has been the discharge of petroleum products into the environment. There are some figures which indicate that recreational boats add as much petroleum and petroleum waste products to the environment as all other sources combines.

Construction

The last source of pollutants listed in Table 10.2. has been construction. By construction has been meant both that done at sea such as offshore oil rigs, and than done on land contiguous with the sea, such as port facilities. Implied hare has been both the actual construction of the structure and the associated activities that go along with construction. There must be all sorts of support equipment, but most important, there must be support people. As in general, the amount of pollution gets directly related to numbers of people, any large construction effort will bring with it the usual people-waste one had not previously has been not too well known, but one things is sure—the environment will never be the same as it was before. Some of these changes might be beneficial, but no matter how beneficial there have been always going to be some members of society who react to any change by claiming a degradation of the environment. Any marine construction project must get judged on the basis of the benefits to be derived by society as a whole rather than solely to those accruing to some small group.

Effects of Marine Pollution

In this article, we shall attempt to get some idea as to what pollution does and why we consider it to be so undesirable. There have been three basic types of pollutants; the pathogenic, the aesthetic, and the ecomorphic. Pathogenic pollutants have been those which cause disease. This disease may be fatal if the pollutant is a lethal poison. Aesthetic pollutants have been those pollutants causing a change in the environment displeasing to the eye, ear, or nose of man.

Ecomorphic pollutants, on the other hand, have been those pollutants which produce a change in the physical characteristics of the environment in such a way that there may be drastic changes in the structure or composition of the biosphere. Obviously, there three types of pollutants have been not of equal importance. Pathogenic pollutants have been certainly much more serious than the other two; however, ecomorphic pollutants are often an indicator of more serious types of pollution to follow. Furthermore, since pollution has been man-produced and the effects have been suffered by man, aesthetic problems are certainly of interest.

Acute Effects

The pathogenic pollutants have been obviously the most important. The effects of pathogenic pollutants are either acute or chronic. The acute effects have been easiest to determine because of the short time between administration and affliction. The easiest way of measuring acute effect has been by feeding the pollutant in question to test organisms, such as mice or selected fish, and observing the dosage required to kill 50 percent of the organisms involved. This is usually carried out by feeding a population, and for each dosage the number of individuals that succumb is noted. When these data are plotted, an S-shaped curve results. Another way of examining acute effect is to observe the effect of a given pollutant concentration on organisms over a somewhat longer period of time, noting how long it takes for a given dosage to produce mortalities. This may be done by giving a predetermined concentration of the pollutant of the test animals and observing how long it takes to kill half of them. When data from an experiment such as this are plotted, a curve similar to vertical asymptote results.

Chronic Effects

There appears to be a dose size for each pathogen below which the pollutant does not have any effect. The concern with threshold values for various pollutants has been one aspect of the study of chronic effects. Chronic effects are the most difficult to measure because of the time involved; in some cases cancers have developed as long as 25 years after initial exposure to the carcinogenic agent. Nevertheless, measurements are made and for many materials a threshold value seems to describe the data reasonably well. Most experiments require some measure of the effect on the organism other than death, so that many different measures are used.

These are; deformity in the growth of organisms, damage to particular organs, change in heartbeat or breathing rate, genetic damage to individuals (rather than to the population as a whole), change in the rate of increase of a population, life expectancy of individuals within a particular population, and fecundity of females within the population. Unfortunately, an index may be very descriptive of a pollutant on a particular species, but it may be completely ineffective in describing what happens when this pollutant comes in contact with other species.

Synergism

Another complication in the examination and quantification of pathogenic pollutants has been the phenomenon of synergism. A synergistic effect takes place when the combined effects of two or more materials acting together has been greater than would be expected from the simple sum of the individual effects. Temperature, for example, has been a very strong synergistic parameter, generally enhancing most pollutant effects as temperature is increased. The same has been true of salinity and dissolved oxygen. Some pollutants, on the other hand, tend to decrease the effects of others, as might be in the case of a strong acid a strong base present at the same time.

Other pollutants may make strong pathogens to get precipitated out of the water column or to be collected on sediments, so that in either case they end up on the bottom. Thus it has been extremely important in the analysis of any pollutant to be aware of the other material present and their synergistic tendencies. Within the marine biosphere many organisms get affected by various pollutants. Whether or not a particular organism has been affected by any given pollutant will depend on many variables. For example, many pollutants occur only in the water column while others, such as some of the pesticides and heavy metals, are adsorbed into suspended sediments and usually find their way to the bottom within a relatively short period of time.

Consequently, these latter types would tend to affect those organisms living on the bottom or feeding from bottom organisms more than they would influence the organisms living within the water column, like free swimming fish. Also of importance in determining the effectiveness of a particular pollutant on living organisms has been its relative solubility in water and oil. Pollutants more soluble in oil will tend to find their way to organisms that have a larger content of oil in their body tissues. This has been particularly true of some of the pesticides which are more soluble in oil than water.

The individual habits of marine organisms also influence their susceptibility to a particular pollutant. Feeding habits with respect to time of day, portion of water column foraged, type of material ingested, and the method of digestion utilized all determine to a great extent the types of pollutants the organism will be exposed to. Similarly spawning habits decide the type of area and time of year in which a species has been particularly sensitive to subtle changes in the environment. These changes influence not only the parents but also the offspring and many be even their progeny. Other habits not having to do with spawning also help to determine whether an organism will get exposed to particular pollutants. Whether the organism has been sensitive to light or sound, for example, will often determine whether it gets attracted to a particular out-fall. Thus we find that a pure and simple cause and effect relationship between a pollutant and an organism has been not enough to determine completely the actual response of the organism in the real world.

Pathways to Man

Once some of the pollutant gets ingested by marine organisms, the major concern of man has been the possibility of this pollutant appearing on his dinner table and causing a pathogenic response. Obviously if a fish ingests a pollutant and a human eats that fish, then that person will get ingested a dose of that pollutant and there will probably be some danger. The question is, "How much?" In order to arrive at an approximation of the actual amount of undesired material finding its way into the reader's stomach, one must be able to follow the pollutant through a number of steps. Most organisms living in the sea tend to concentrate materials existing in the sea to much greater values than they have been in the ocean itself. For example, a diatom, a small form of marine algae having a silicate frustule, is having a body concentration of silicon about 40 thousand times as great as the oceanic waters from which the organism derives all of its material. Thus the diatom concentrates silicon very effectively so that if we are diatoms, a large portion of our diet would be glassy.

Similarly, other organisms, including fish, will concentrate other materials even if these do not naturally occur in the environment. But this amplification or concentration does not stop with one step. We must continually keep in mind the fact that an organism does not exist by itself in the marine environment. Any organism has been dependent upon many other organisms for its existence. An edible fish, for example, might very well subsist on smaller fish which, in turn, might subsist on small zooplankton (small floating forms of animal life), which in turn, may subsist on phytoplankton (forms of floating plant life). There may be 5 to 10 individual links in a food chain leading to man, and there has been the distinct possibility that for each one of these links a concentration of the pollutant occurs. Hence, the total pathway to man for the pollutant might be a rather tortuous one, but it might result in a very high concentration of the pollutant material in the fish. This itself may be of no concern to the individual unless he ingests the fish and if so, how often and how much he consumes.

It has been not enough simply to know that there has been some pollutant present in marine organisms; one should also know how much is present, and in addition two other facts have been required. This first has been the accepted maximum level of ingestion of the pollutant before harm results, and the second has been the amount of fish that can be eaten before this level gets reached. In this way it might very well be possible to have a particular kind of fish one or two times a year with perfect safety, while if it has been eaten once a day, it might very well have a toxic effect.

Ecomorphic Pollutants

There are many different types of pollutants which may be classified as ecomorphic, while some pollutants fall into all three of the classes stipulated above and many fall into at least two. Sediment has been a pollutant whose effect has been primarily that of changing the environment since it is not pathogenic and very often is not aesthetically displeasing, although it can be. As it settles to the bottom, it will tend to bury the organisms that live on the bottom, such as oysters and clams, and eventually smother them. Sediment also tends to fill in marsh areas, killing the marsh grass, and completely changing the habitat. As marsh areas are breeding grounds for small fish and an important ecological link in most of the oceanic life, any destructive process such as this has been bound to markedly change the marine biological environment. On the other hand, in some cases where man's activities are able to produce large amounts of erosion and sediment has been brought down by rivers into estuaries, some estuarine areas have been filled enough to support marsh lands rather than destroy them.

Whenever there has been a depth change in an estuarine or coastal area, the effects have been many. They range from changing the biological environment so that larger fish will no longer live in the shoal areas to affecting the total circulation pattern of the area. If the sediment has been in suspension it also has an effect on the ecosystem, the most obvious aspect being a change in transparency. By making the water more opaque, light will not penetrate as deeply as before so that the thickness of the layer of water in which plants grow (the euphotic zone) will get decreased. If the water has been fairly shallow to begin with, this decrease in transparency may very well result in the inability of rooted plants to grow and some marsh lands may be destroyed.

The effect of sediment, then, has been two-fold: changing both bottom and water properties. Both of these effects tend to change the physical characteristics of the ecosystem. Another ecomorphic pollutant has been oil, although oil sometimes produces toxic effects. The result of oil coating bird feathers has been well known physical effect. The oil does not poison birds; it simply makes it impossible for them to fly, and consequently many of them die. A film of oil on the surface will also decrease the amount of sunlight entering the water and, similarly, will limit the exchange of oxygen from the atmosphere.

In addition to the aesthetic and pathogenic effects oil, there are thus also ecomorphic effects. The toxic effects of various nitrogen compounds have been well known but fertilizers and other nutrients washed into the sea possess a nontoxic effect on marine waters. These materials will do just what they have been designed to do: make plants grow, whether these plants be terrestial or aquatic plants. Consequently, the first effect of a spike of fertilizer added to the sea has been an overstimulation of those plants normally there. Beyond that the undesirable plants (those that are not food for the edible types of marine animals) often get stimulated to grow and these marine weeds grow so rapidly that the overwhelm the other types of plants.

As plants have been at the bottom of the food pyramid, the structure and characteristics of the ecosystem have been completely changed, even to the extent of markedly altering the species composition. Once these plants have started to grow at greatly accelerated rates, they may completely cover the surface of the water, disallowing any light from penetrating more than a few centimetres. Those plants on the bottom of the blanket will die, due to lack of light, and those rotting plants will use oxygen in their decay processes. Even though there have been more plants than before, producing more oxygen in the process of photosynthesis, so many of them are dying and decaying the result has been to decrease the dissolved oxygen. This phenomenon is termed as eutrophication. Solid waste has been another type of pollutant very often disposed of in oceanic areas.

In some cases they tend to completely destroy the natural habitat and the result has been not only ecomorphic but also aesthetically displeasing. However, in some cases, the ocean dumping of waste products such as old automobile tires and even automobile bodies has produced an artificial habitat for many organisms that were not endemic to the region previously. However, there are cases where nontoxic solid waste materials have destroyed desirable habitats. Sometimes the dumping of solid wastes makes a change in water flow patterns producing changes in salinity and temperature which influence the biosphere markedly. This usually takes place when the water is reasonably shallow and the volume of solid waste products dumped is relatively large. Another ecomorphic pollutant that seems to be receiving a lot of attention lately is heat which is supplied to the marine environment from electrical power plants or other industrial sources. This heat will increase the water temperature and may result in different kinds of damage to marine organisms. As most mutations have been caused by thermal activity under normal conditions, the possibility of genetic damage has been real.

It is estimated that about 90 percent of the mutations occurring normally have been thermally caused, while only increase in the temperature, one can expect greater mutation rates or more genetic damage. In addition to genetic damage, heat can cause both acute and chronic problems. Animals can be killed by extreme thermal shock or they can suffer lingering effects and finally expire by being unable to acclimate to the new surroundings. More subtle changes are also noticed as growth rates as changed with increase or decrease in temperature. Normally a small increase in temperature will make a small increase in growth rate; however, if the temperature change is extreme, this may be reversed.

Changes in spawning patterns have also been noted in which fish that normally move upstream to spawn are unable or unwilling to cross a thermal barrier associated with a heated effluent. This may cause the cessation of spawning for large portion of the species population. Another effect of temperature increase has been to cause certain organisms to grow more rapidly than others, upsetting the previous ecological balance and eventually causing a change in species composition. A similar range of effects to that produced by heat has been caused by salinity changes. These might be produced from the input of rich brine solutions resulting from the effluent of water desalination efforts or a fresh water outfall such as a raw sewage treatment plant. A marked change in salinity will bring about damage similar to that produced by heat, including genetic damage growth rate changes, spawning pattern changes, and changes in species composition. All organisms are used to living within some range of salinity and if this range gets exceeded, they will either die or move some place else.

Aesthetic Pollutants

The last type of pollutant mentioned above was aesthetic pollutants. Aesthetic pollutants have been those which offend the human senses and as such as regarded by many to be unimportant. In many environmental problem areas, though, the aesthetic considerations often receive more attention by environmentally oriented citizens than some of the pathogenic pollutants. This is simply because the aesthetic pollutants have been visible and what can be seen, smelled, or heard appears much more bothersome than that which is hidden, even though the hidden pollutant might be much more dangerous to life. Consequently, any scheme for controlling marine pollution must include some consideration of the aesthetic pollutants if it is to succeed.

An excellent strategy seems to be to focus on the aesthetic pollutants and use them as a mechanism to gather momentum to clean up the rest of the pollution problems. Solid waste material, when piled in shallow areas where it can be seen, has been a typical aesthetic pollutant. Building activities also may be regarded to be sources of aesthetic pollutants. Power transmission cable has caused more than one power company grief because of the environmental consideration given the local citizens who are concerned about the appearance of power cables running across the countryside.

Oil wells, specially the offshore variety visible from resort areas, have been of great concern to most environmentalists. The same thing has been true of equipment required for dredging operations and the spoil areas resulting from these operations. To some people, especially those in living relatively pristine environments near the water, boats can also be an aesthetic pollutant, especially those that cause waves and noise. Effluents that make the colour of the sea to change or increase the amount of suspended sediments also may be regarded aesthetic pollutants. It simply has been not as pleasant to go swimming in water that is cloudy or contains colouring material different from what one would expect in a normal, natural body of water. In essence, any change from the pristine environment appears to many to be aesthetic pollution.

Obviously, we are living in an unreal world if we think the marine environment can be maintained in an unspoiled form. It should be kept in mind that many complaints about aesthetic pollution come from individuals who would rather live in the primeval environment not realizing that it would not be available to them unless the commercial world about them supported it. There appears to be a dichotomy wherein we try to maintain the unspoiled environment in certain areas while letting others become over utilized. Nevertheless, there has been a growing movement to include in the cost of every project a relatively small sum to improve its aesthetic qualities. This is, of course, exactly the same philosophy that has been applied to the control the pollution-producing properties of the effluent because society believes that this additional sum of money has been a desirable expenditure. It may be similarly be maintained that the expenditure for aesthetic pollution control is desirable as it also will tend to increase the quality of life.

Control of Marine Pollution

Now that the nature of pollution and some of the sources of this undesirable product of society have been considered, we will now concern ourselves with some of the avenues open to us to accomplish some sort of control. The most obvious control strategy that can be used has been to somehow or other completely prevent the introduction of the pollutant into environment. This can be carried out by not producing the pollutant or by restructuring the process os that the pollutant is no longer a by-product. Let us consider a few major sources of pollutants and some possible methods of control in these particular cases.

Dredging

Maintenance dredging, that is, dredging required to maintain a particular depth, adds sediment to the water column and would not be needed if there were no sediment supply to shoal up major channel areas. This type of dredging has been not a one-shot affair, but it has been almost always necessary to periodically redredge the accumulated sediment from these channels and harbor areas. This continual sedimentation has been a normal process which occurs whether man is present or not.

In some instances the activities of man tend to accelerate the sedimentation process but this has been not necessarily always true. For example, coastal plain estuaries, often the result of river basin flooding, tend to be a rather ephemeral event in geologic time. They get formed when water level rises, and they disappear with the continual process of deposition of material eroded from the higher reaches of the water shed. Another example of natural erosion has been found on the west coast of the United States in the westernmost portion of California (that portion where more than 90 percent of the population lives). This area was formed by erosion of the Rocky Mountains and deposition of the eroded material at the base of the mountains. The most obvious solution to the problem of maintenance dredging has been to cut off the sediment supply to the areas where sedimentation has been a major problem. This can be carried out in a number of different ways. The most effective of all has been to prevent upstream erosion, but at the same time this is perhaps the most difficult.

The entire watershed must be policed very carefully to make sure that the natural landscape gets changed to such an extent that erosion no longer takes place. This implies extensive terracing and replanting of areas that have been denuded for one reason or another. Of course, this also implies that extreme measures must be taken in areas that do not have natural cover, such as agricultural areas or areas where construction is underway. A certain amount of this is being done. Another method of preventing sediment from reaching the undesired area has been to provide settling ponds. The water is released after it has dropped the major portion of its load and is allowed to proceed downstream.

The major disadvantage of this system has been that it is essentially a temporary solution since the ponds will fill up rather rapidly, needing either building of new ponds or dredging of some sort. Sometimes the sedimentation that takes place in a channel area is a result of sediment loads that are introduced into the marine environment from nearby rather than those carried from far upstream. This takes place for example, when a channel gets dredged with unstable sides so that the material just tends to slough into the deeper regions. This needs continual dredging as the removed sediment has been almost immediately replaced by bottom material from the sides.

One method of correcting this situation is by stabilization of the bottom. Adding larger, more dense material such as gravel to the bottom has been a technique utilized occasionally, although this has been usually more expensive than dredging so that it has been not done too often. Bottom stabilization, however, is a technique that is being used more and more to prevent the movement of bottom materials within relatively small contiguous areas. Another possibility for preventing the accumulation of sediments in undesirable areas has been to change the natural circulation patterns so that sediments get deposited in an area where they do no damage. This may be carried out in various ways: by changing the physical dimensions of estuarine areas, by putting dams in selected areas, and even by changing the circulation pattern by planting grasses to increase bottom friction.

Another method to be considered in decreasing the amount of suspended sediments has been that involved in controlling the numbers and types of water craft allowed in particular areas. In regions that have been relatively shallow and narrow, high speed vessels tend to churn up the bottom, putting sediments back into the water column that had previously settled out. In addition, large wakes are generally produced that tend to erode the sides of the waterway by the mechanical action of these relatively high energy waves breaking on the shoreline. Both these effects can get decreased by the simple expedient of limiting the speed of boats in certain areas.

It has been a fairly effective method; however, if the traffic is extremely heavy it may be necessary to limit the total number of vessels, in addition to their speed. In addition to the general clarity aspect of pollution caused by dredging, there has been also a toxic one. Sediment retrieved from harbor bottoms has been especially likely to contain a relatively high concentration of industrial pollutants such as oil and heavy metals. During the dredging operation there has been a likelihood of some of this material's finding its way back into the water column, but of more concern has been the longtime leaching from dredge spoil dumping areas. Various methods, such as diking and the construction of artificial islands, are utilized to minimize this leaching but their performance under all conditions has been not too well known. Further research in the handling of dredge spoil gets needed at this time.

Oil Pollution

Another problem that, to a certain extent, appears amenable to prevention at the source is that involving oil. Unfortunately the sources of oil in the marine environment have been extremely difficult to delicate. All the experts agree that the greatest terrestrial source of petroleum products in the ocean is used crankcase oil. Some go so far as to indicate that this source supplies more petroleum to the sea than any other source. This may be true because within the last decate or so the habits of American motorists are changed markedly to the extent that the a large portion of them now change their own oil, discarding the used oil in a manner that probably makes it to end up in storm sewers.

The major reason for this probably has been the almost complete disappearance of the used lubricating oil re-refining industry in recent years since it was not economically feasible to re-refine oil. However, with the modern continual increase in the price of new oil, the re-refining industry has been having a rebirth so that a large portion of the oil previously thrown away will now be reused. Another entirely different source of oil in the marine environment has been that resulting from accidents that take place during shipping and transferring processes. Generally speaking, the more handling steps involved, the more accidents and associated losses can get expected. The figures as to how much has been actually lost in handling accidents compared to that lost in normal ship procedures, such as pumping bilges, has been not well known, but there has been no doubt that a very large portion of the oil reaching the marine environment has been the result of poor handling procedures in both the operation of vessels and the transfer of oil from one carrier to another.

Better procedures, stricter enforcement of accepted safe procedures would go a long way to decreasing the amount of oil that enters the ocean each year. When a tanker empties its load of oil, the dynamic characteristics of this vessel get markedly changed. Its weight gets decreased about a thousand times and therefore it becomes extremely difficult to driven the ship, because it was designed to be driven with a full load. Therefore, when most tankers do not have load of oil, they will fill their tanks with water for ballast. When they return for a fresh cargo of oil, they must pump this water from the tanks, releasing large amounts of oil as the residue left in the tanks gets washed out.

There are many schemes which are suggested for decreasing the amount of oil that is wasted in the ballasting-unballasting process; however, it is not clear at this time how effective these processes have been or how effectively any requirement can be enforced. Modern supertankers have been extremely large vessels and therefore have major structural weaknesses, so that occasionally they will break up in unusually heavy seas. They have been simply not designed to be utilized under unusual storm conditions, so that there will continue to be losses of this type in rough weather. In passing, it should be mentioned that once oil gets spilled at sea under large wave conditions (usually the case when accidents occur), there has been no way for this oil to be retrieved.

On the other hand, oil may be retriered in harbor areas where the sea surface is relatively clam, but only when there are waves no higher than approximately 1 metre. The obvious way—the only way—to prevent oil from getting to the marine environment when a ship founders has been to prevent the accident from occurring in the first place. Alternatively, once the accident has taken place the oil must be prevented from escaping from the ship. It has been somewhat doubtful whether these solutions will ever be possible.

Industrial Pollutants

Most industries are using large amount of water for cooling purposes or as an integral part of the manufacturing process. Consequently, industrial effluents may be having waste products, heat, leached material from heat exchangers, or even incidental house-cleaning wastes. If the industry, for example, has been using water as a coolant, the heat exchangers must constantly be cleaned to keep the heat exchange efficiency high. These cleaners may be more toxic than process wastes. Consequently, it has been extremely desirable to treat all industrial effluents. Some industries have tried to modify the entire process rather than just clean up the effluent. This often causes an increase in efficiency with a lower overall production cost. If this expenditure can be recouped in a relatively short period of time, this obviously might be expected to be a very cost-effective method of handling unwanted effluents. However, in many cases, there has been no increase in opening efficiency, with the net result being an increase in manufacturing costs passed on to the consumer. Society must then make a decision as to how much it has been willing to pay for a particular product vis a vis the amount of pollution introduced in the manufacture of this product.

Antifouling Paint

Sometimes pollution comes from the use of antifouling paint on boats. During the summertime in temperate waters the growth rate of marine organisms get increased markedly, especially those organisms setting on fixed surfaces. An untreated boat hull left in the water for a period of just a few weeks will need enough growth to cut the available speed for a given power input by fifty percent. Consequently, all owners of boats left in the water have been very careful to apply some coating to the underside to prevent the accumulation of marine organisms. For large cargo ships or tankers, this could make savings of thousands of dollars in fuel costs. Common practice has been to paint the bottom of the boat with antifouling paint. Antifouling paint has been so used to leach a toxic material into the water in just the proper amounts so that all organisms setting on the hull will get killed. However, the older types of antifouling paints, those that use copper as a toxin, released enough material to the environment not only to kill possible fouling organisms but also to be harmful to other marine creatures at some distance from the boat. Thus there has been a movement afoot to improve the quality of antifouling paints by drastically reducing this leaching to toxic material into the water.

The solution that has been suggested has been to use new paint vehicles in conjunction with new toxic materials so that the toxic material has been almost completely retained within the vehicle, requiring organisms to come within a very short distance of the boat to be affected. One result of recent research is to develop a family of paints using acrylic resins as vehicles and tri-butal tin fluoride (TBTF) as the primary antifouling ingredient. In the first few years of use this paint holds great promise because it is less toxic in the water, lasts longer, and does not allow bottom growth more effectively. In their very effectiveness these new paints pose a problem, because the toxin gets retained within the paint for a longer period of time.

Paint scrapings have been now more toxic than they were with the old copper bottom paint. This means that it has been somewhat more dangerous to strip the paint from the bottom of the boat when applying a new coat and precautions must be used to prevent the old paint from getting back into the water environment. When sandblasting large ships, for example, precautions have to be taken to retain the sandblasted material rather than allowing it to be simply washed away. Thus the solution of one problem presents another, but these problems have been solvable if some care has been taken to see that additional health hazards do not get created.

Pesticide and Herbicide Pollution

Some of the juicier pollution scandals that have taken place in recent years have involved various pesticides and the problem has been that many of the chemical compounds get introduced for public sale before the total affects of these materials have been known. It seems to be almost an insurmountable problem to determine what the total effects of a new compound will be before it has been put on the market because so many new products are introduced very year. However, a large portion of the problem has been undoubtedly due to basic carelessness.

In both the manufacturing and use of these materials proper precautions are not being taken into consideration. Many users, for example, undoubtedly believe that if a certain amount per unit area will suffice, twice as much will produce twice as much control of either weeds or bugs. This excess toxin is washed into the streams and ends up killing marine organisms rather than protecting plants. It would be evident that more effective control of both the manufacture and use of pesticides and herbicides would go a long way toward eliminating undesirable side effects.

Solid Waste Pollution

The disposal of solid waste has been very necessary and there even seems to be a fear that if the solid waste problem is not solved, eventually there will be no room left for people. Because of this fear, large amounts of solid waste get dumped in the ocean. Some of these solid wastes take the form of sludge from sewage treatment plants whereas others consist of large pieces of discarded consumer products.

In the latter case the problem might very well solve itself as the cost of scarp metal continues to increase. One possible use for sewage treatment sludge has been in refurbishing land disfigured by strip mining. Often the overburden removed to get at the coal seam has been not capable of supporting life nor it has been even particularly stable. The addition of sludge from sewage treatment plants is considered as a method of soil treatment to make it possible to support a viable plant population. It is a concept that has not been tried yet but one which will be tried when it becomes economically feasible to transport the sludge to the strip mine.

Another possibility for using the sewage sludge, if it contains only nutrient materials, has been suggested and has been considered by some of the Middle Eastern oil producing countries. The idea has been that empty tankers be filled with sewage sludge for the return voyage rather than with ballast water. This sludge will then be utilized to bring the desert regions back to a point where they have been capable of supporting a large agricultural industry. Other forms of solid waste being dumped in near-shore areas have been dumped there primarily because there has been no other acceptable depository.

One method of decreasing the severity of this problem has been simply to decrease the amount of solid waste produced by society. Packaging of consumer items, for example, seems to be a good place to start. From a disposal point of view the catsup battle is probably the worst possible package. The package weighs more and takes up more space than the contents. The Navy has done excellent work in decreasing package size and weight because there are definite limitations abroad ship, but not too must has been done with the private sector of society simply because there hasn't been enough action on the part of consumers to insist upon small volume and weight packages.

One possible exception to this has been the introduction of lightweight plastic battles for large size drinks to replace the heavy weight glass variety. They are somewhat thinner than glass, weigh less, and may be incinerated, so the disposal problem has been somewhat alleviated. However, with the continuing increase in petroleum costs, plastics might not remain economically feasible in the future.

Recycling and Reclamation

It has been found that biological communities living in the sea need a certain amount of physical shelter. Some species just like to be alone, whereas in other cases the young of the species require places to hide to protect against predation. A marsh or wetlands area has been a good breeding place because both of these conditions prevail. It is found that in many cases these conditions can be created artificially by introducing large junk articles such as old automobiles or worn out ships to form effective artificial fishing reefs. There has been disadvantages to using waste products for this sort of activity. One has been that in many cases these junk objects get transported to undesirable areas by dynamic oceanic forces.

Pollution has been not controlled when automobile fenders and doors wash up on swimming beaches. When creating a finishing reef, steps must be taken to retain the integrity of these reefs for a long period of time. If proper attention has been paid to the magnitude and nature of oceanic forces, artificial reefs might be made relatively permanent. With the present state of the economy where things are some what more expensive, it appears that more items will find themselves in the near future on the list of materials to be recycled or reclaimed.

Industrial processes will alter as it become more feasible to extract materials from the waste stream from a cost standpoint and, at the same time industrial, waste streams will become less and less polluted. Even in the area of multiple waste product use, like the combination of waste heat from either manufacturing or electrical power generation and waste nutrients from sewage treatment plants for the purpose of aquaculture, cost will always be the controlling factor. It is known that elevated temperatures and the artificial addition of nutrients will increase growth rates of many organisms markedly. The choice of which organism to grow has been involved not only with which would be best to grow, but also which there has been a market for.

Setting up an aquaculture system involves picking organisms which respond favourably to the environment, developing a system by which the optimum growth rate can get produced, developing a marketing system so that the end product will be sold at the maximum possible price, and assuring the continuing existence of a market for everything produced. Not only has been science included but also sociology and economics must be considered, and it is probably in these latter two areas that the major difficulties reside. Hence, no sophisticated aquaculture systems employing both waste heat and nutrients have been developed at this point, but it has been probably only a matter of time before we will have extensive aquaculture using man's waste products.

Storage of Pollutants

Another method of controlling pollution has been is to store them at some distance from man's activities so that they have been as inaccessible as possible to the major fraction of the population and pathways to man. Occasionally nuclear waste products are being dumped in deep oceanic areas in the hope that when they do break out of their canisters, they will not find an easy pathway back to man because the deeper life forms have been probably not in any food chain involving man. Perhaps not so well known has been the fact that deep ocean disposal is used for other materials such as overage ordnance and poison gases for which the military has no further use. In the case of nuclear products, there has been a very vociferous school of thought that believes storage facilities should allow waste retrieval if a technological breakthrough is ever accomplished that would allow these nuclear waste products to be profitably utilized.

Asimov's suggestion has been to implant our hazardous waste products in the oceanic trench areas so that within a few thousand years they will get carried into the interior of the earth and no longer be bothersome. The only problem with this suggestion has been that the exposure time before the canister having the waste products is effectively buried in the crust is relatively long. During this time the canister seal has to absolute. Whether this can be accomplished with present technology is presently being debated within the scientific community. Nevertheless, storage has been still a viable method of controlling pollution, especially for the more toxic materials. At least theoretically, one knows where they are. The major concern with storage of toxic material has been accidental release, and the changes of this taking place always remain within the realm of possibility.

Controlling Pollution by Zoning

Another method of controlling pollution has been by specifying certain areas that can be utilized for effluent discharge and other areas that cannot. Although we are even starting to convert our city dumps into usable areas, such as golf courses or ski slopes by covering, planting and land-scaping as rapidly as possible, but it will be some time before ocean dumping areas are reclaimed in this fashion. Certain coastal regions would be reserved for industry, others for sewage treatment, while still others would be set aside for recreation. Some of these activities have been such that they may get pursued in conjunction with one another, whereas others cannot.

The idea has been for every coastal area to try to get the optimum utilization of marine resources available. In order to accomplish this type of zoning most effectively, a great deal more knowledge is needed than is presently available about the synergistic effects of various types of activities. One of the major advantages of optimum zoning has been that it would allow maximum avoidance of undesirable synergistic effects. The biggest stumbling block to coastal zoning probably has been inovercoming entrenched interests in regions where the optimum activities have been found to be different from those presently underway: a socioeconomic problem rather than a technological one. The concept is some what interesting one, because it essentially permits some pollution of certain areas while refusing to permit any pollution of other locals. This has been different from the way pollution has been presently controlled, where the effort has been made to limit pollution in all areas to the minimum possible amount. With the zoning concept pollution would still be limited as much as possible, but it might permissible to allow the water quality of ocean to degrade below the mean in some places, because in others it would be higher than the mean.

Controlling Pollution by Taxation

One last method of pollution control is taxation which can be utilized to encourage or discourage certain activities in addition to raising funds to support government programmes. Pollution would be allowed, but it would be taxed at a rate proportional to the environmental insult. Some economists suggest that a tax on pollution will simply be a license to pollute and will therefore encourage pollution rather than discourage it. From a management point of view there have been a number of different directions that can be taken to control pollution, including both technological and socio-economic alternatives. So far society has tended to aim more in the direction of the technological alternatives, but there appears to be a definite swing toward other methods. In the years to come it might be possible that more and more emphasis will be kept on these nontechnical methods, especially as the trade-offs, costs, and benefits become better understood.

11

CONTROLLING MARINE POLLUTION

Environment Effect of Oil and the Chemical Used to Control it

There are many factors, acting both individually and in combination, which govern the effect of oil on marine life. These will include the type and quantity of oil, the direction and quantity of pollution experienced by the organism, the state of the oil, e.g., fresh, weathered etc., the season, with respect to the annual cycle of the organism, and the habitat and the natural stresses to which the organism is subjected. Any or all of these can affect the organism in many different ways. The effect may be lethal or sub-lethal, in the latter it can disrupt physiological or behavioural activities and death may result due to interference with feeding of reproduction. The uptake of oil can result in tainting or possible carcinogenesis either of the organism contaminated or one higher up the food chain. The visible and well-known effects are beach pollution and the death, due to oiling, of seabirds. It is not unreasonable to wonder, if not actually to fear, whether oil might nor adversely affect marine life and, in particular, harm fish the source of much food for man and his domestic animals either by reducing their numbers of rendering them unfit to ear. Like all life, life in the sea depends upon solar radiation for its source of energy.

Through photosynthesis, marine plants, of which by far the most important are the minute drifting organisms collectively know as phytoplankton, store this energy in a form which other organisms such as zooplankton and the fish which prey upon it can utilize, a simplified diagram showing the board outline of what is in reality a highly complex food web. It is helpful to realize that at each step the efficiency of conversion of food into body weight is only about 10%.

Nelson Smith estimates that man would require at least on kilogram of phytoplankton to gain one gram in weight. This is a useful statistic for it highlights the fact that direct effects on fish stocks eaten by man are of much greater economic importance than equivalent damage in terms of biomass lost, to the planktonic organisms on which the fish depend. The food web diagram also highlights the fact that fish, and other organisms well removed from the primary production stage represented by the phytoplankton, have by proxity, consumed relatively enormous quantities of organic materials some of which, although only present in small concentrations in phytoplankton, may collect at the higher tropic levels of the food web. The question of whether there is an accumulation of petroleum hydrocarbon compounds, if they represent a credible threat to the resource or the consumer, and the other possible effects on marine life of oil and of chemicals and methods used to clean it up.

Plankton

At almost all major oil spills, lay men and ecologist alike express fears about possible harmful effects on plankton. Concern can usually be summarised under three headings:

- A layer of oil on the surface will, by occluding light, cause a serious drop in photosynthesis by phytoplankton.
- The oil layer will interfere with gaseout interchange at the sea surface and dissolved oxygen levels will thereby be lowered.
- Oil and dispersants will exert serious toxic effects on plankton.

How real are these effects and what is their ecological significance? The first and second are the indirect consequences of measurable physicochemical phenomena and will be treated first.

Occlusion of light

Ottway studied the light transmitting properties of 20 crude oils made up a 5% solutions in white sprit and showed that different crude oils vary widely in this respect. Oils high in asphaltenes and sulphur were found to absorb most light and, for all oils, the absorption spectra exhibited maximum at the blue end of the spectrum (325-350 nm), indicating that the 44 same group of compounds is mainly responsible in all of the oils tested. A number of photosynthetic pigments found in phytoplankton also hive absorption maxima at the blue end of the spectrum. Potentially, at least, their photosynthetic activity could be impeded by crude oils spilt at sea. It is important to remember the scale of the problem. A huge slick of black oil covering 2000 square miles of the North Sea actually covers less than 1% of its area. It is safe to conclude that occlusion of light by spilled oil, It phenomenon easily demonstrated in the laboratory, has no ecological significance for the plankton of the open area.

Effects on dissolved oxygen levels

The solubility of oxygen in most hydrocarbon mixtures is similar to that in sea water. An oil slick is Dot, therefore, an impenetrable barrier to gaseous interchange. For a full discussion of this issue, the reader should consult Nelson -Smith; suffice it to say that in a 2 day trial, Brown and Reid found that build water, beneath a 17 mm layer of oil, absorbed 73% as much oxygen as the control. Confirmation that oil slicks do not interfere significantly with gaseous interchange at the surface of the open sea is provided, by an observation by the Scripps Institution of Occeanography, on dissolved oxygen levels below a heavy oil slick in the Santa Barbara Channel. The oxygen saturation beneath the slick was 98.5% of that in water outside the slick area.

Toxic effects

There is abundance evidence that freshly-spilled crude oils contain low boiling point substances which are actually toxic to the planktonic organisms. The most toxic of these are water soluble aromatic derivatives. Low molecular weight alkanes, (10 carbon atoms or less) may induce narcosis in high concentrations but these are unlikely to result from a crude oil spill in the open sea. The aromatic solvents used in early emulsifiers were also found to be highly toxic to planktonic organisms at levels of less than one part per million. In letter formulations of equivalent dispersing power, the use of non-aromatic solvents has reduced toxicity by over two orders of magnitude. Useful as laboratory assessments of toxicity can be for comparing the acute toxicity of oils and dispersant chemicals to small number of organisms they are rarely a reliable guide to effects on wild populations. In that such tests are necessarily short term they under estimate toxicity, but on the other hand, the toxicant concentrations used in laboratory experiments are rarely maintained for any length of time in the sea. Evaporate losses the rapid dilution of residual, soluble compounds ensures that toxic effects are short lasting, even for high toxicity, solvent emulsion of the aromatic type. Thus although highly localised short-term effects on plankton were observed following the massive use of high concentrations to disperse Torrey canyon oil.

There was no evident that any planktonic species, was significantly affected the population level. The Torrey canyon experience was quite exceptional in the amount and toxicity of the materials used to dispense the oil, and even these it would seem that the rapid evaporation and dilution of the toxicity combined with the disposed nature and rapid turnover of the plankton to minimize the ecological impact. Por other major spillages, including the notorious Santa Barbara blow-out the toxic effects on the planktonic tissues have proved to be virtually detectable. The plankton of milfod Haven was examined to see if the development of a major oil port had any effect. The report stated that changes which have occurred are no greater than might have been produced by natural environment changes and population dynamics. Sanbourn suggests however that the effect on benthic larvae when they arc in upper layer of the ocean may be more longer lasting as benthic animals required years to reach maturity and so the effect on the larvae may endour longer.

The only planktonic species which may conceively be in any ecologically significant danger from the toxicants associated with crude oil or dispersants are those which form fairly localised aggregations. Arrow worms fish eggs, and larvae may come in to this category. The likely hood of ecologically significant effects in such cases is .examined in detail. Finally, to keep our assessment of toxic effect on plankton in their true perspective it is helpful to consider the scale of plankton production of in the sea relative to the that of petroleum spilled. Fraser estimates annual world phytoplankton production to be about 150,000 million tons and that of zooplankton 15,000 million tons. By contrast estimates of annual losses of petroleum in the vary from 2.5 million to an improbable maximum of to million tons. Only a fraction of this petroleum is toxic to plankton and very little of this is ever likely to come into contact with planktonic organisms.

Fisheries

As was said earlier, the purpose of this chapter is to arrive at an assessment of the effect of oil pollution, an environment stress, for which man is largely responsible, on the biology of the sea. It is important, in this context, to consider the particular case of sea fisheries, a sector of the marine environment in which man already exerts a direct and measurable influence. The term, 'fisheries' has a variety of meanings, the widest embracing all exploited stocks of marine organisms from macro-algae to whales, and the narrowest exploited species of fish alone. In this chapter, the term is taken to mean the fish and shellfish of commerce. Oil pollution is potential treat to these resources for three main resources for three main reasons.

- Pollutant any directly harm fish and shellfish stocks.
- Pollutant adhering to, or accumulated by, fish and shellfish may render them unacceptable to the consumer.
- The presence of spilled oil may interfere physically with fishing operation.

Controlling Pollution

Now that the nature of pollution and some of the sources of this undesirable product of society have been considered, let us concern ourselves with some of the avenues open to us to accomplish some sort of control. The most obvious control strategy that can be used is to some how or other completely prevent the introduction of the pollutant into the environment. This can be done by not producing the pollutant or by restructuring the process so that the pollutant is no longer a by-product. Let us consider a few major sources of pollutants and some possible methods of control in these particular cases. The most obvious solution to the problem of maintenance redging is to cut off the sediment supply to the areas where sedimentation is a major problem. This can be done in a number of different ways.

The most effective of all is to prevent upstream erosion, but at the same time this is perhaps the most difficult. The entire watershed must be policed very carefully to make sure that the natural landscape is change to such an extent that erosion no longer occurs. This means extensive terracing and replanting of areas that have been denuded for on reason or another. Of course, this also means that extreme measures must be taken in areas that have no natural cover, such as agricultural areas or areas where construction is under way. A certain amount of this is being done and these are a number of legal requirements for persons engaged in construction or agricultural activities. Another method of preventing sediment from reaching the undesired area is to provide settling ponds. If a stream is identified that carries a large sediment load, a pond and dam arrangement can be built so that the velocity of the stream is decreased to zero and the suspended sediment will have a chance to settle.

The water is released after it has dropped the major portion of its load and is allowed to proceed downstream The solution since the ponds will fill up rather rapidly, requiring either building of new ponds or dredging of some sort. Sometimes the sedimentation that occurs is channel area is a result of sediment loads that are introduced into the marine environment from nearby rather than those carried from far upstream. This occurs, for example, when a channel is dredged with unstable sides so that the material just tends to slough into the deeper regions. This requires continual dredging as removed sediment is almost immediately replaced by bottom material from the sides. One method of correcting this situation is by stabilization of the bottom. Adding larger, more dense material such as gravel to the bottom is a technique utilized occasionally, although this is usually more expensive than dredging so that it is not done too often. Bottom stabilization, however, is a technique that is being used more and more to prevent the movement of bottom materials within relatively small contiguous areas.

Oil Pollution

Another problem that, to a certain extent, appears amenable to prevention at the source is that involving oil. Unfortunately, the sources of oil in the marine environment are extremely difficult to delicate. Many figures have been published and, in almost all cases, there is a significant disagreement as to the amount of oil discharge into the marine environment and the sources of the oil. There is no doubt, though, that a large portion of this amount originates on land as waste product. As would be expected, the major source of this is associated with the automobile. Some of this lost oil is dropped from leaks in cars and trucks, some is deliberately dumped as used crankcase oil while some results from the deposition of unburned exhaust gases. From roads and streets this material is washed into the marine environment, usually without benefit of treatment.

Even when passed through treatment plants, however, petroleum-rich effluents still provide problems as they are difficult to treat and they tend to make treatment for other pollutants less effective. All the experts agree that the greatest terrestial source of petroleum products in the ocean is used crankcase oil. Some go so far as to indicate that this source supplies more petroleum to the sea than any other source. This may be true because within the last decade or so the habits of American motorist have change markedly to the extent that a large portion of them now change their own oil, discarding the used oil in a manner that probably causes it to end up in storm sewers. The major reason for this probably is the almost complete disappearance of the used lubricating oil re-refining industry in recent years since it was not economically feasible to refined oil. However, with the modern continual increase in the price of new oil the re-refining industry is having a rebirth so that a large portion of the oil previously thrown away will now be reused. Another entirely different source of oil in the marine environment is that resulting from accidents that occur during shipping and transferring processes generally speaking.

The more handling steps involved, the more accidents and associated losses can be expected. The figures as to how much is actually lost in handling accidents compared to that lost in normal ship procedures, such as pumping bilges, is not well known, but there is not doubt that a very large portion of the oil reaching the marine environment is the result of poor handling procedures in both the operation of vessel and the transfer of oil from one carrier to another. Better procedures are need, but perhaps even more than better procedures, structure enforcement of accepted safe procedures would go a long way to decreasing the amount of oil that enters the ocean each year. When a tanker emplies its load of oil, the dynamic characteristics of this vessel are markedly changed. Its weight is decreased about a thousand times and therefore it becomes extremely difficult to drive the ship, since it was designed to be driven with a thousand. Therefore when most tankers do not have a load of oil, they will fill their tanks with water for ballast. When they return for a fresh carg of oil, they must pump this water from the tanks, releasing large amounts of oil as the residue left in the tanks is washed out.

There have been many schemes suggested for decreasing the amount of oil that is wasted in the ballasting-unballasting process; however, it is not clear at this time how effective these processes are or how effectively any requirements can be enforced. Modern supertankers are extremely large vessels and therefore have major structural weaknesses so that occasionally they will break up in unusually heavy seas. They are simply not designed to be utilized under unusual storm conditions, so that there will continue to be loses of this type in rough weather. In passing, it should be mentioned that once oil is spilled at sea under large wave conditions there is no way for this oil to be retrieved.

On the other hand, oil may be retrieved inharbor areas where the sea surface is relatively calm, but only when there are waves no higher than approximately one meter. The obvious way the only way to prevent oil from getting to the marine environment when a ship founders is to prevent the accident from occurring in the first place. Alternatively, once the accident has occurred the oil must be prevented from escaping from the ship. It is somewhat doubtful whether these solutions will ever be possible.

Recycling and Reclamation

If it is impossible for one reasons or another to prevent the introduction of a pollutant into the system, it may be possible at some place along the line to extract this pollutant and rescue it. It should be kept in mind that although "consumer" is a commonly used word, there is no such thing as a consumer. The word tends to imply that material is being used up, while, in actuality, it is just borrowed for a while and returned. It may be in a somewhat different form or perhaps a somewhat different shape, but nevertheless it is still returned. We don't destroy any of this mass; we just change its form, so that the general concept of recycling or reclamation is extremely appealing especially when we consider our larger solid forms of waste. If we could find some way to utilize all of these materials, we would solve a number of our problems especially those involved with space. Sometimes, though, it is not necessarily desirable to solve a space problem sometimes the space problem works to our advantage. For example, it has been found that biological communities living in the sea require a certain amount of physical shelter.

Some species just like to be along, whereas in other cases the young of the species require places to hide to protect against predation. A marsh or wetlands area is a good breeding place because both of these conditions prevail. It has been found that in many cases these conditions can be created artificially by introducing large jack articles such as old auto mobiles or worm out ships to form effective artificial fishing reefs. There are disadvantages to using waste products for this sort of activity. One is that in many cases these junk objects are transported to undesirable areas by dynamic oceanic, forces.

Pollution is not controlled when automobile fenders and doors wash up on swimming beaches. When creating a fishing reef, steps must be taken to retain the integrity of these reefs for a long period of time. If proper attention is paid to the magnitude and nature of oceanic forces, artificial reefs can be made relatively permanent. In actuality the whole recycling and reclamation business is predicated on cost. If it is economically feasible to recycle or reclaim material it will be recycled or reclaimed. If it costs more money to use the recycled material than new material than the material will not be recycled.

It is difficult to imagine that certain materials such as glass, will ever be recycled to any great extent because silicon is one of the most common elements in the earth's crust and it costs just about as much to make a new bottle as it does to recycle an old one. However, in the case of aluminium there is about a factor of five involved in recycling coast as opposed to manufacturing from scratch. With the present state of the economy where things become more expensive from year to year, it appears that more items will find themselves on the list of materials to be recycled or reclaimed.

Industrial processes will change as it becomes more feasible to extra materials for the waste stream from a cost standpoint and, at the same time, industrial waste streams will become less and less polluted. Even in the area of multiple waste product use, such as the combination of water heat from either manufacturing or electrical power generation and waste nutrients from sewage treatment plants for the purposes of aquaculture, cost always be the controlling factor. It has been known for some time that evaluated temperatures and the artificial addition of nutrients will increase growth rates of many organisms markedly. The choice of which organism to grow is involved not only with which would be best to grow, but also which there is a market for.

Setting up an aquaculture system involves picking organisms which respond favourably to the environment, developing a system by which the optimum growth rate can be produced, developing a marketing system so that the end product will be sold at the maximum possible price, and assuring the continuing existence of market for everything produced. Not only is science included but also sociology and economics must be considered, and it is probably in these latter two areas that the major difficulties reside. Consequently, no sophisticated aquaculture systems using both waste heat and nutrients have been developed at this point, but it is probably only a matter of time before we will have extensive aquaculture using man's waste products.

Storage of Pollutants

Another method of controlling pollution, utilized for certain pollutants, is to store them at some distance from man's activities so that they are as inaccessible as possible to the major fraction of the population and pathways to man. Thus we essentially throw them away, but we thrown them away, in a place where no one can get to them and their effects cannot get to man. Occasionally nuclear waste products have been dumped in deep oceanic areas in the hope that when they do break out of their canisters, they will not find can easy pathway back to man since the deeper life forms arc probably not in any food chain involving man. Perhaps not so well known is the (act that deep ocean disposal has been used for other materials such as average ordiance and poison gases for which the military has not further use. In the case of nuclear products, there is a very vociferous school of thought that believes storage facilities should allow waste retrieval if a technological break through is even accomplished that would allow these nuclear waste products to be profitably utilized. Another school feels that they should be placed in an areas as completely inaccessible as possible to everyone. This latter school intuitively seems the safest, but, on the other hand, most common storage places are just not the type.

There is always a small probability of a pathway back in man being found. One suggestion for a non-retrievable storage place which holds some promise has been made by Issac Asimov. According to modern Geological theory the earth's crust is composed of a number of separate plates, all moving with respect to one another. In some places the plates are moving together and in others parts following the motion of the mantle, the layer of material underneath the earth's crust. As a result of convection, the mantle tends to rise toward the earth's crust in certain areas, producing oceanic ridges such as the North Atlantic ridge.

At these divergence areas, where the mantie material is brought up to the earth's crust, the motion becomes horizontal and the crustal plates are carried away from the ridges at a rate of a few centimeters per year. The convergent regions where the plates come together are where the mantle material is carried back down toward the centre of the earth. In direct contrast to the regions of divergence associated with ridges, the convergent areas are associated with oceanic trenches where crustal material is carried down and under the continental plates. These oceanic trench areas the very deep, often as deep as 10,000 meters, and at the same time are actively engaged in transporting material into the bowels of the earth.

Control Pollution by Zone

Another method of controlling pollution is by specifying certain areas that can be utilized for effluent discharge and other areas that cannot. This method has been used on hand as records have been kept. Although zoning of certain regions as industrial, commercial, or residential is a relatively recent innovation, there always was city dump. Now-a-days we are even starting to convert our city dumps into usable areas, such as gulf courses or sky slopes by covering, planting and landscaping as rapidly as possible. This is an exciting concept that is gaining as rapidly as possible. This is an exciting concept that is gaining is popularly as dumping areas become more and more scarce, but it will be some time before ocean dumping areas reclaimed in the fashion. Even though we are attempting to utilize our dumping areas for other purposes as rapidly as we can, the concept of zoning still pervades our society. Certain areas are set aside for certain types of use. Society is willing to accept a factory in a particular location where it is hidden, while at the same time people will not accept that factory as a next door neighbor.

The same concept has been suggested for marine coastal areas. Certain regions would be reserved for industry, others for sewage treatment, while still others would be set aside for recreation. Some of these activities are such that they may be pursued in conjunction with one another, whereas others cannot. The idea is for every coastal area to try to obtain utilization of marine resources available. In order to accomplish this type of zoning most effectively, a great deal more knowledge, is required than is presently available about the synergistic effects of various types of activities. One of the major advantage of optimum zoning is that it would allow maximum avoidance of undesirable synergistic effects.

The biggest stumbling block to coastal zoning probably is in overcoming entrenched interests in regions where the optimum activities are different from those presently underway: A socioeconomic problem rather than a technological one, the Coastal Zone Management Act encourages coastal states to attempt to take the first step in planning for optimum utilization of the coastal areas. After plans are complete, the states area to implement them with the financial assistance of the federal Government. The concept is an interesting one, because it essentially allows some pollution of certain areas while refusing to permit any pollution of other locales. This is somewhat different from the way pollution is presently controlled, where the effort is made to limit pollution in all areas to the minimum possible amount. With the zoning concept pollution would till be limited as much as possible, but it would be permissible to allow the water quality to degrade below the mean in some places, since in others it would be higher than the mean.

Controlling Pollution by Taxation

One last method of pollution control that has been suggested, although it has met with a great deal of resistance and antagonism, taxation. As is well known, taxation can be utilized to encourage or discourage certain activities in addition to raising funds to support government programs. Pollution would be allowed, but it would be taxed at a rate proportional to the environmental insult. This general philosophy is followed at the present time by requiring the polluters pay fines roughly proportional to the amount of pollution, and in many cases industries will purposely pollute and pay the fine because it is apparently cheaper than making the necessary plant modification required for cleanup. Obviously the system is not working if this is the case;

the fines are too low. Some economists suggest that a tax on pollution will simply be a license to pollute and will therefore encourage pollution rather than discourage it.

Other economists, however, indicate that pollution taxes should be based on the severity of the pollution with some sort of graduated tax. A little pollution requires a little tax, a large pollution requires a large tax. An important aspect of this concept is that the words the true and total cost to society of these pollutants. Every time the common resource is utilized by an individual, this society for the right to use this common resource. This payment would be in the form of a tax set at a level determined by how much society economists believe that anyone utilizing the resource should pay for its degraded then the individual degrading should pay for the degradation. This is an interesting concept that is being discussed in economic circles taxation certainly is a powerful tool for societal behaviour modification that has worked many times.

Generally speaking, taxation encourages particular activities and discourages other if the taxation amount is large enough to completely discourage undesirable activities. From a management point of view there are a number of different directions that can be taken to control pollution, including both technological and socioeconomic alternatives. So far society has tenced to aim more in the direction of the technological alternatives, but there appears to be definite swing toward other methods. In the year to come it appears that more and more emphasis will be placed on these nontechnical methods especially as the tradeoffs, costs, and benefits become better understood.

The Forces Involved in Dispersing Pollutants

Once a pollutant has been introduced into the marine environment, it becomes a matter of no small interest to determine where this pollutant goes and how the concentration of the pollutant varies with time and location. These process fall into two basic groups : advection and dilution, advection is the mechanism by which the pollutant is carried from one place to another, while dilution results when the pollutant is caused to reduce its concentration by some sort of spreading mechanism. Advection simply has to do with oceanic currents and will include three primary types of current since these are encountered most often in the oceanic regime.

Management of the Marine Environment

Up to this time we have been primarily concerned with the technical aspects of pollution. Space has been devoted to discussion of what pollution is, where it comes from, and where it goes in the marine environment. Now we must consider how to control this pollution so that the problems resulting from it are kept to a minimum. The previous sentence has a number of words in it which should probably be defined very carefully, but probably never will be. For example, the statement that pollution should be controlled so that the resulting problems be kept to a minimum may be interpreted many different ways be many different people since a problem for one group is not necessarily a problem for clear-cut. Just as one man's trash is another man's treasure, one man's pollution is another man's enrichment, and this is not only dichotomy that faces a conscientious marine manager. There are problems of jurisdiction involved in whether the responsibility for running the operation lies with the federal government, a local government, or some regularly commission, and always there is the application of priorities supposedly set by society as a whole. The basic tool the manager has to work with is the law, however, and this is where we begin.

The Legal Background

The legal background for pollution control goes back to the Rivers and Harbors Act of 1899, recently interpreted so as to be used as a pollution control measure. The act set up responsibilities for protecting navigational and environmental aspects of all navigable waters to the U.S. Army Corps of Engineers. The Corps therefore has the responsibility not only of protecting navigable waters so they remain navigable but also protecting them against their loss as a viable vehicle for various other uses that society considers deriable. An interesting aspect of this law and one reason it has become such a powerful tool in the hands of the environmental protectors is the redefinition of navigable waters. It turns out that the modern definition includes just about every stream going into a large river in the United States. Thus the Army Corps of Engineers was given the responsibility for protecting the environment 1899 and to all intents and purposes still retains the primary responsibility as far as brackish and salt water is concerned.

In 1934 the Fish and Wildlife Coordination Act was passed This has since been amended many times, but essentially it involves the fact the U.S. Fish and Wildlife Service and The National Marine Fishers Services have the responsibility for determining the effect of any proposed project on fish and wildlife resources. They then report to the Corps of Engineers which makes the final decision as to whether or not the proposed project is allowed. In 1948 the Federal Water

Pollution Control Act was passed and this also has been emended periodically. It requires the Corps of Engineers to apply environmental standards and criteria in approving the disposal of spoil in navigable waters.

The most recent amendment of this act was the Federal Pollution Water Control Act of 1972. This was a complete rewrite of the 1948 law and it set out for basic requirements. In the first place, a national goal for the elimination of all pollution from America's waters by 1985 was proclaimed. This is the famous zero discharge goal Note that this is not a mandate to have zero discharge by 1985, but as a national goal it is something to aim at and it is to he hoped that we will come as close as possible. Second, the Act of 1972 set as another goal the addition of secondary treatment to all·municipal sewage treatment plants by 1977. This goal has not been met but there has been a strong effort in that direction. Third was the effort to encourage the use of new, more advanced treatment and disposal methods by 1983 so that some of the problems will be resolved by the use of advanced technology rather than by simply improving in present methods. And last, the act established an industrial cleanup program with restrictions becoming tighter with the passage of time.

Each year the requirements involving effluent standards will be tightened so that these will be continuing cleanup of the nation's waters. This particular aspect of the bill was backed by stiff penalties including both fines and imprisonments. This 1972 Act is the basic law under which most antipollution suits are brought. The Water Pollution Control Act of 1948 encouraged the creation of uniform laws to control pollution, supported research, or firms to cease practices leading to pollution, and established the Federal Water Pollution Control Advisory Board. This has been succeeded in recent years by the Council Environmental Quality. The year 1965 saw the passage of the Water Quality Act, In this act states were given an opportunity to adopt and enforce federal water quality standards for interstate waters, The Clean Water Restoration Act of 1966 authorized massive federal participation in the construction of sewage treatment plants about 3.5 billion dollars were allocated to be spent between 1967 and 1971.

The National Environmental Policy Act of 1969 provided preparation of environmental impact statements when the U.S. Army Corps of Engineers determined it to be necessary. The Act also declared it to be national policy to encourage the production of enjoyable harmony between man and his environment. The Water Quality Act of 1970 addressed oil pollution from vessels and offshore facilities, federal permits, sewage pollution from vessels and hazardous substance discharged into the nation's waters. Also in 1970 reorganization plan No.3 was submitted to the Congress by the Chief Executive. This reorganization plan established the Environmental Protection Agency, setting up the organization and transferring responsibility for various activities from other agencies Unfortunately, the executive order does not address the problems of the objective or the purpose of the Environmental Protection Agency (EPA) and this objectify or purpose has been left to the Agency Administrator to interpret for himself. This omission has resulted in a general grey area with EPA being as strong or as weak as the Administrator and Congress want it to be.

In 1972, in addition to the amendment to the Water Pollution Control Ad, two other act of some import to the protection of the marine environment were passed. The first of these was the ocean Dumping Act of 1972. This act requires the Army Corps of Engineers to apply Environmental Protection Agency standards and criteria in approving ocean disposal of dredge spoil materials where transport passes through United States territorial waters. This is an interesting law because it controls the material that originates in United States ports and not that material which is damped in United States water. Consequently no matter where the material is dumped, as long as it originates in a United States port, the Corps of Engineers has some control over it. The other act pass in 1972 having important ramifications to the marine community was the Coastal Zone Management Act.

This act gave the individual states primary responsibility for determining land and water uses to more effectively balance environmental protection add economic development objectives in the coastal zone. It delineated the coastal zone as including all navigable waters and defined them as all waters up to the mean high tide line, including we lands wholly or partially covered at high tide, whether privately or publicly owned. Under this act large amounts of money have been given to participating states, and all coastal and Great Lake states have so far indicated a desire to participate. Funds are initially for planning purposes aimed at developing a general zoning plan for the use of the coastal zone later funds will be available for the implementation of these plans. In an attempt to get at specific causes of pollution, the Ports and Waterways.

Safety Act was passed in 1972 which addressed the problem of pollution from ships. This act allowed for ship design and construction regulation with respect to possible environmental damage and it applied to all United States flag vessel as well as all foreign vessel entering

United States waters carrying, either oil or hazardous liquid cargo in bulk. Vessel design and operational standards for all these vessel in United States waters were finally promulgated on 23 January, 1977. In this manner the United States government is attempting to prevent as many large catastrophes involving vessel carrying oil or hazardous materials as it possibly can by simply not allowing marginal vessel to navigate within the territorial limits of the United States. Another fact of marine pollution having a source outside the continental area is involved with driliving for oil and gas offshore. The Outer Continental Shel Lands Act gave the responsibility of Granting leases for continental shelf drilling to the Department Interior. These leases required the leases to reimburse the federal government for any oil spill cleanup cost and in addition they require the installation of bolwout prevent equipment on the wells themselves.

International Convenants

The foregoing is a brief list of the basic framework of federal laws which govern the discharge of waste products into our terristorial waters. In addition there are a few international convenants under which the United States operates. The Intergovernment Maritime Consultative Organization (IMCO) has sponsored three conventions resulting in a number of regulations; The first of these was the 1954 Oil pollution convention which was amended in 1962 and 1969. This convention places limitations on the rate of discharge, the oil content, and the distance from land for oily ballast discharge water from ships. The enforcement of this convention is left up to the flat state, that is the country where the ship is registered. The second convention was the 1971 amendment to the Oil pollution Convention which branched out in different direction by addressing vessel design.

Unfortunately this part of this amendment has not been rectified by any maritime nation. In 1973 a ship pollution convention was sponsored by IMCO, and this convention resulted in a set of regulations establishing standards for ship board sewage treatment equipment and also specified the requirement for segregate ballast tanks for new ships of specified sizes. It also specified methods for storing dirty oil ballast residue, oil discharge monitoring system, and oily water oil extraction systems. Unfortunately this convention also has not been ratified by any martime nation.

Consequently, behaviour on the high seas is left strictly to the conscience of the individual ship's master. At this time there are also no international agreements on ocean dumping or the environmental effects of seabed resource exploitation. There have been many attempts to draw up an International Law of the Sea agreement but as of this writing none has been successful.

Jurisdictional Responsibility

One of the problems facing a marine resource manager is the problem of jurisdictional dispute. Who does have responsibility? In the four general areas the responsibility is somewhat dispersed. The first of these areas is water drainage, capture, or use which is exclusively on a local level. Since this is primarily a problem of private rights which is usually involved with state of private ownership. This area does sometimes get involved as a pollution source so it is of concern to us. The third area of interest includes the bays and coastal waters which are controlled by a combination of state and federal regulations. Generally speaking, the state has control over resources and the federal government has control over navigational and environmental considerations, but there are many exceptions. Most state laws are not quite as stringent as the federal environmental requirements so that the federal requirements must be obeyed.

However, there are some state laws more strict than the federal and when this is the case, the state laws take precedence. The last areas is the region beyond the coastal region wherein the responsibility is somewhat loose. We have seen that there are no viable international agreements for the deep ocean, but in many cases the United States has acted unilaterally to protect its continuous waters. When the passage of the Ports and Waterways Safety Act and the Outer Continental Shelf Lands Act attempts have been made to protect against the effects of oil and hazardous material, cargo ships and offshore oil-wells.

Political Processes

Once the problem of jurisdiction is settled, there then appears the problem of political processes. The political process is such that technical standards will almost always yield to such things as hardship cases, emergency situations, or strong public sentiments. Consequently, any decision to be made where the environment is involved must have a public input or else the decision will probably not be effective. The manager must not only face; an often emotional public, but also individual agencies, both state and federal, working at cross purposes. For example, the Environmental Protection Agency is primarily concerned with water quality, the U.S. Fish and Wildlife Service is primarily concerned with fish habitat improvement, while the Army Corps of Engineers has a primary concern in the areas of navigation improvement. These concerns must be balanced against common public needs such as high employment, protection of scenic environments, and housing demands, and very often these conflict very strongly so that at least one of the many groups must eventually compromise.

The compromise, in the long run, must be based on some sort of a set of societal objectives. Society must somehow or other determine the most important things it wants from its marine resources. The activities range from aesthetic enjoyment of the area to commercial shipping and waste disposal. Obviously waste disposal is not exactly the type of activity to enhance aesthetic enjoyment, nor is utilizing the coastal zone for military maneuvers encouraging to the maintenance of a refuge or a sanctuary. Some of the uses of the marine coastal area must be sacrificed in certain cases for other uses which society deems more important. Somehow or other society must set priorities so, that the areas can be managed to obtain the ends that society wants from them. This is a very important point and one that is often completely ignored.

In very few cases has society overtly set down priorities for the use of marine areas, and until priorities are specially stated, marine area managers will be forced into making had decisions. It is to be hoped that some of the planning being done under the Costal Zone Management Act will involve priority setting, especially regarding the use of particular areas. If this is the only outcome of the coastal zone management act, it will indeed be a valuable contribution to the management of the activities of society in the marine coastal zone.

Costs, Risks, and Benefits

In order to set priorities in a rational manner, managers must by one means or another determine benefits, cost, and risks associated with any and all activities. One these three parameters have been quantified, then cost/benefited rations may be determined and priorities may be set on the basis of these ratios. However, this is much easier said than done. All three of these parameters are usually quite qualitative and very difficult to quantity. Benefits, for example, are often what people perceive them to be and therefore are essentially a political evaluation rather than an exact calculation. Costs can often be determined accurately in terms of the actual physical nature of the activity. Buildings, machinery, and labors can be quantified in terms of their cost. However, very often an attempt to control pollution or the effect of pollution on society results in costs that simply cannot easily be quantified.

Risks also are difficult to evaluate. How much is a life worth? Does it make a difference if it is yours or someone else? Thus we find that we must be satisfied with some approximation to these three parameters. Some methods for obtaining approximations to values for, costs risks and benefits will be discussed below. The cost versus pollution control curve, is an exponential curve wherein the initial costs of controlling pollution are relatively inexpensive, but the finals 5 or 10% of pollution control may cost as much as the initial 90 to 95%.

The benefit pollution control curve, on the other hand, is parabolic in shape. Here the benefits in the initial cleanup phase are quite large while the benefits accrued from the final stages are quite small. We might consider then some sort of a cost/benefit ratio, knowing .these two functions. The optimism cost benefit ration is determined on the basis of the incremental cost per incremental benefit. If example, an additional dollar of environmental control will produce more than a dollar's worth of benefits, then it is more effective to spend additional money on control. If, however, additional control results in fewer dollars worth of benefit, then it appears that the control process has been carried too far. The optimal level of pollution control is not at point A where the cost of pollution control and benefits are the same, but is at point B, where the rate of increase of cost per unit pollution control is equal to the rate of increase of benefit per unit pollution control. At point B the slopes of the Curves are the same. We may then write

$$\frac{dc}{d(pc)} = \frac{db}{d(pc)}$$

From a cost benefit analysis then it appears that it is not desirable to completely clean up the environment since the benefit secured would not be worth the expenditure required. However, this is not the complete story. It is obvious that in certain cases even a small amount of pollutant is not acceptable because of the danger to health. Thus the risk too much be considered. Each activity and each pollutant has an associated risk, some of these to the environment and some to man. It is difficult to quantify these but one method commonly used is to determine the number of fatalities per million hours of exposure for each person involved in the activity considered. This is expressed as a statistical probability. Table 11.1 is a listing of risks expressed in this manner for various common activities.

Notice that the activities are divided into two groups, voluntary and involuntary activities. The voluntary activities arc those which the individual can decide for himself whether or not he wishes to do.

Notice that many of these activities have a probability very close to that of death by disease, including old age. Others, however, such as rock climbing, have an extremely high risk associated with the them and therefore are only for the adventuresome. Interestingly enough, the risk associated with the one involuntary activity listed, electric power, is less by about three orders of magnitude than most of the risks in the voluntary portion of the table.

Table 11.1. Risk Table

Risk	*Death rate per million hours of exposure*
Voluntary	40.0
Rock climbing	6.6
Motorcycle	2.4
Scheduled airline	2.4
Smoking cigarettes	1.2
Death by disease	-
Including old age	1.0
Private auto	0.95
Sking	0.71
Railroad and bus	0.08
Involuntary	—
Electric power	0.002

Apparently then we are loath to let others do in to us what we happily do to ourselves. Pollution, of course, falls within this area of involuntary activities since any pollutant discharged to the marine environment is not directly controllable by any individual. Consequently we would expect that the risks associated with pollutants would be down in the area of that of electric power rather than up with that of railroad and bus travel. In order to quantify priorities, it if necessary then to develop some sort of risk benefit ratio, which in turn, requires quantification of benefits.

One possibility of quantifying benefits is to simply assume that in the case of voluntary activities the amount of money spent on the activity by the average involved individual was proportional to its benefit, while in the case of involuntary activities, the contribution of the activity to the individual's annual income as proportional to its benefit. Here risk is plotted relative to the benefit for various kinds of voluntary and in voluntary exposure. From these graphs and some other works that has been done, a number of conclusions can be drawn.

First, the public is apparently willing to accept voluntary risks roughly a thousand times greater than those involved with involuntary exposure. Second, the statistical risk of death from disease and old age appears to be a psychological yardstick for establishing the level of acceptability of other risks, Third, the acceptability of risks appear to be crudely proportional to the third power of the benefits, whether they be real or imagined. Fourth, the social acceptance of risks appears to be directly influenced by public awareness of the benefits of an activity as determined perhaps by advertising or the number of enhanced by increased public awareness. A determination of the level of optimum pollution control and general management schemes involved with addressing the total pollution problem is not as easy one. We find that everything costs something.

There is no such thing as a free lunch For every cleanup process involved there is a cost. Sometimes this cost is minimal but other times the cost is extremely high (For example, it has been estimated that in order to meet the 1983 goals of the Federal Water Pollution Act or 1972 it would cost 350 billion dollars just to treat municipal sewage in an adequate manner. This includes the cost of controlling nonpoint sources along with sewage treatment plant updating. Whether or not the America public is willing to spend this much on this sort of program is at this time unclear, but the American public must be made a ware of the necessity of costs and trades of any cleanup program. If we are going to spend this amount of money for pollution cleanup then there will not be as much money available for other activities, and once again we must decide which to these activities is more important to us. But the decision can only be made if we have as much data as possible in hand to allow as accurate as possible a determination of benefits, costs, and risk involved with each process.

The Future of the Marine Environment

In the previous pages we have seen that the oceanic volume is extremely large, while at the same time, that portion of the oceanic volume available for pollutant discharge is relatively small. Consequently, we hope to have demonstrated a need for some kind of marine pollution management since there is a very real problem associated with dumping wastes in the ocean. It has also been shown that an ever increasing population, distributed primarily around coastal areas, will inevitably increase the total amount of waste discharged to the oceans. This is coupled with the fact that technology is continually growing and its growth will be associated with an increasing amount of pollution per capita with the passage of time. It all boils down to the realization that the total volume of water available for waste dissolution will remain relatively constant while the total amount of waste will probably increase as far into the future as can be foreseen at this time.

Management Requirements

Fortunately, however, in those nations which have reached a relatively high level of technology, there is an increasing awareness of the problem and a public requirement that planning and management proceed in order to utilize the marine, resource in the best manner possible. In order to do this a number of requirements must be met. In the first place the effects and pathways of all pollutants dumped into the ocean must be known. This is an extremely difficult task. Many of the pathways are known but the effects are not. For other pollutants the effects are well known but the pathways back to man are still mysterious. This obviously will require a great deal of research, and even then there is no guarantee that all effects and pathways for all pollutant materials will ever be known.

Second, the average residence time for all materials discharged into the marine environment should be known. Here again we are faced with an almost impossible problem due to the tremendous amounts and number of different materials discharged into the oceanic environment. There should also be a continuing monitoring of the environment if only to determine gross changes so some action may be taken before the situation gets out of hand. Lastly, we must continually determine and redetermine what people consider to be important. The more and folk ways of society constantly change, altering values and priorities.

Setting Priorities

At this time managers have no clear understanding as to what people's priorities are, nor is then: any real indication that the people themselves have any great insight into their needs and desires. We do not know, for example, whether the major portion of the populace considers sport fishing or ocean dumping to be higher in priority with respect to usage of the marine environment. We do know what the small special interest groups have to say about this situation, but in terms of jobs lost or consumer goods price increases which would result from the discontinuance of ocean dumping, there is very little information. In order to determine society's needs and desires the people must first be educated as to the costs, benefits, and risks associated with any environmental protection solution being considered.

Obviously a much better choice of alternatives, can be made if the advantages and disadvantages may very well be a utopian dream, but it certainly should be the goal of each manager to come as close to this as possible. At the present time choices are made on the basis of limited knowledge since not all of the costs, risks, and benefits are available to the manager making a decision, every those that are available to the manager are usually not available to the public at large so that the manager's decision can be evaluated. This situation must change if utter choice is to be avoided. Once our cards are on the table where they may be seen by all interested parties, we may proceed by setting goals that are both popular and attainable. A goal such as "zero discharge by 1985" is not an attainable one and when such a goal is set and not met, very often the result is a certain cynicism regarding all future goals.

Goals, Guidelines, and Standards

At this point it may be desirable to define some words commonly used in the management of the marine environment. In particular there are four words needing clarification: goal, guideline, criterion, and standard. Although the definitions which follow may not necessarily agree with those in a standard dictionary, they follow the usage of most people involved in marine are management, standard : A plan established by governmental authority as a program for water pollution prevention and abatement.

Criterion: A scientific requirement on which a decision or judgment may be based concerning the suitability of water quality to support a designated use.

Guideline: An acceptable methodology for achieving any given standard.

Goal: The ultimate standard which all current standards may eventually approach.

Involving the Public

Thus goals must be set with two criteria in mind. One is the scientific requirement defined above, but there is also a personal requirement on the part of the public based on its sense of priority. The public will not support a particular criterion, even with substantiating scientific data, if it interferes markedly with the present life-style. Until the citizenry can be educated to the risks involved in not accepting scientifically based criteria, looser criteria must surface. But it is the duty of the manager not only to choose the acceptable criteria but also to attempt to educate so that more logical criteria may receive public support. Along with the choice of particular goals, standards, criteria, and guidelines for a particular time, it must be assumed that these are not permanent but may very well change with the passage of time. Both the environment and public attitudes must be constantly monitored to effectively meet the needs of a changing society.

With a proper monitoring system, not only may the standards, criteria, and guidelines be change but also the goals may update. In this manner the public is made a working part of the management process adding support to the process and consequently making it more effective. The coastal Zone Management Act was designed along these line and it has been in operation long, enough for a major portion of coastal states to have prepared detailed plans on the optimum use of coastal areas. These plans are undergoing extensive public hearings and it is to be, hoped that when they are finally adopted the public will actively support them. If this is indeed the case, it can be hoped that the effect will, at the very least, be to stop the increase in marine pollution and may be even reverse the trend by decreasing pollution. In any event, however, regulations set down to protect the environmental quality of the marine environment will be enforced with popular support.

Impact Associated with Cleanup Techniques

Cleanup Effects on Marshes

The equipment used for several cleanup techniques considered here does not impact on the marsh itself. For example, mechanical and weir skimmer are deployed in tidal channels, and pools, not in the marsh proper. Vacuum skimming in itself produce no impact on the marsh, but it is frequently accompanied by substantial foot traffic. Fool and equipment traffic and considered secondary impacts and may, in themselves, produce substantial adverse effects. The following sections are modified from Maiero et at. (1978).

Effects of Traffic

The most obvious type of disturbance caused by traffic in a marsh is physical breakage of plants. In both grass dominated and succulent-dominated marshed, physical damage effectively reduces the amount of photosynthetic tissue and may expose the interior of the plants to toxic fractions of the oil. Plants are likely to recover from one such event, but recurrent trampling may clear a part that will persist for years. If the soil is soft, as it is in many marshes, roots and rhizomes may be broken and thereby accelerate erosive processes. Foot traffic potentially accelerates erosion even where the substratum is firm and the rhizome may remains essentially intact.

Shorelines that consist of steep escapements are particularly vulnerable. Cleanup activities that entail foot traffic should be used in such regions only after other methodologies have been considered and rejected. Sometimes it is necessary to transport heavy equipment to remote portions of marsh. The impacts of foot traffic described above apply even more strongly to vehicular traffic. Traffic on soft marsh soil may bury plant stems and leaves and reduce their productivity until they resprout or grow back above the soil surface. Further, traffic under these conditions is likely to bury oil. In the anerobicsoils that characterize mangroves and marshes, residual oil may persist for years. If the buried oil is toxic, it may inhibit the growth of anything in the contaminated zone until it eventually dissipates.

Flushing

Low pressure flushing with sea water from a nearby source and thus likely to be of a salinity to which the marsh is accustomed may be beneficial if used with caution. Where vegetation cover is continuous and sediments relatively stable, low pressure flushing may be effective in removing substantial amounts of oil from vegetation. Drawbacks of the method are:

- All oil is not removed from leaves and a sufficient amount may adhere to the waxy plant cuticles to cause damage. This is especially true when the oil is fresh in which case it adsorbs onto leaves very strongly an penetrates them.
- Foot traffic required to deploy the flushing equipment may cause damage.
- Oil flushed to creeks, etc., will be available to re-oil the marsh or other areas unless properly collected or other wise removed.

Where vegetation cover is incomplete where much bare mud is in evident, or where sediments are sort and muddy, sediment disturbance or erosion may produce additional damage to that of the oil: Wherever possible, sediment disturbance should be avoided.

High pressure flushing may cause some erosion, local rearrangement of the substrate, and physical damage to the plants. These forms of damage may be less severe than similar impacts caused by foot traffic, but this depends upon mode of deployment. High pressure flushing may came oil to be driven into the substrate. If steam or heated water is used, marsh animals and plants in the spray pattern may be killed or stressed by thermal stock.

Sorbents: Sorbent pads, oil snares, and similar cleanup aids have two major drawbacks:

- They are usually used by a large group of personnel who heavily traffic the marsh and
- They must be recovered. Additionally, if cleanup teams are not trained in the use of these material, there is chance that oil may be mixed with a shallow layer of the substrate in the course of recovery.

Because complete recovery of sorbent materials is seldom possible in a field cleanup exercise, remanants of sorbents may persist. Most remanants are merely unsightly, although some may ensure birds and animals. Large concentrations of under-graded materials could block the light from an appreciable portion of the marsh surface and consequently reduce marsh productivity in that region. Biodegradable sorbents avoid many of these problems. Oil sorbent materials (synthetic and biodegradable) should always be recovered from the marsh, those that are not collected are a potential source of recontamination and a hazard to marsh animals.

Cutting

Cutting of oiled plants is a cleanup technique that entails direct physical destruction of plant tissues. As such, it severely reduces the amount of photosynthetic tissue and may expose the interior of the plant to toxic substances in the oil. Moreover, cutting operations are likely to require a great deal of foot traffic and the attendant adverse effects. Cutting is probably most beneficial with certain species of grasses and rushes that have been heavily contaminated with viscous oil and are not subject to natural cleaning. If the entire aerial portion of the plant is coated, the roots are likely to suffocate unless some passage way is opened to the interior of the plant.

Cutting accomplishes this, provided that free oil that might replug the cut stem has been removed from the surrounding marsh. However, if oiled marsh plants are not thoroughly coated with oil air can still diffuse down the stem to roots, so that cutting is unnecessary unless other threats are present. Spartina marshes are very tolerant of occasional cutting, especially late in the growing season. Saltwort marshes are less tolerant. Burning is sometimes an effective method of removing oil and contaminated vegetation from a marsh without encouraging injurious foot traffic. Spartina marshes can withstand occasional burning.

In fall and winter, they die back to a state of dormancy. During this period, the plants are dry and may support burning. In fact, fall burning of marshes used for agriculture is a commonly applied management tool. During the dieback period, burning can be achieved in spartina marshed without damaging the buried portions of the plants and can stimulate their regrowth. In all other seasons, however, not only in Spartina difficult to burn, but the Browning portions, shoots, and buds are injured by burning. However, this damages does not necessarily permanently harm the marsh since the underground portions of the buried plants are likely to sprout and replace the destroyed portions soon after the burning. Saltwort marshed do not die back seasonally and the plants do not have large, protected underground systems; thus, burning is injurious at any time.

Soil Removal

Removal of soil entails elimination of marsh plant habitat and should be avoided. Nonetheless, if the substrate is heavily saturated with toxic oil and no predicted to recover naturally, this may be the only available cleanup technique. On soft areas of find mud, it may not be feasible at all.

Dispersant Chemicals

Dispersant chemicals have been applied to oil in salt marshes following spills and in a number of experimental situations. In many cases, these have caused additional damage to the oil alone, and in some cases can be demonstrated to enhance the penetration of oil into intertidal sediments. Recent evidence suggests that new generations of dispersant may cause little additional geological effects to oil along when used at appropriate concentrations. However, the value of using them in most cases if unclear particularly on extensive marshes where, following application, dispersed oil may enter creeks to affect other part of the marsh or inertial systems. The may be of value in cleaning small and stable fringing marshed in combination with low pressure flushing, but, in general, dispersant use is not recommended for salt marshes. It should be borne in mind that spraying of near dispersants onto vegetation products damage and correct dilution to manufacturer recommendations is vital. Marsh cleaning methods do not appear to decrease damage done by oil and often increase in. The need for cleanup should be carefully reviewed before any action is taken.

Cleanup Effects on Mangroves

Few publications are available that deal directly with the effects of oil spill cleanup on mangroves. The several observations of the effects cleanup on mangroves at the Florida Keys oil spill of 1975. So many authors summarize effects of cleanup of marine wetlands and include numerous general statements concerning mangroves. The following summary is based on the publications and their personal observations several scientists made during cleanup operations within oiled mangroves.

Recovery of Marshes from Oil Damage

Although a definition of recovery has been attempted, mainly to distinguish the process from that of restoration, the definitive includes the terms structure and function. The majority of scientific literature on the effects of oil deal with changes in the structure of marshes relatively few deal with the function of marshes let alone the part oil damage may play. Hence, we may be able to determine the time at which a marsh returns to something approaching its original structure, but not necessary.

Alternatively, a similar marsh function could be achieved with a different structure. This is a line of investigation that deserves further attention, but at present we can do little but recommend that marsh structure be used in the assessment of recovery. In an earlier section, the rate of recovery of salt marshes has been shown to be considerably influenced by a large number of factors Their precise action and interactions are often poorly known, which means that in practice each spill must be treated as unique and assessment of damage must be dealt with in each case on its own merits little more than generalization is possible from the scientific literature. Recovery processes begin when oil toxicity is reduced or removed.

In most cases, recovery starts irrespective of the activities of man, although cleanup or other action may influence the rate. Where oil damage is relatively light and little of the toxic material remains in the system recovery can be rapid either by regrowth from rootstock or from seeds. Baker reports that single oiling in field experiments, although toxic to many salt marsh species, may produce effects detectable only for one to three growing seasons. Where oiling is repeated or where damage is servere and toxic materials are retained in sediments, recovery may take much longer, and the effects of the spill may be detectable over decades. Where erosion of sediments takes place, recovery may not occur.

OIL SPILL PREVENTION AND CLEANUP IN THE VICINITY OF CORAL REEFS

Prespill Mapping

Prespill contingency planning is now an accepted part of preparation for, and response to, oil spills. Vulnerability mapping has been used by several authors to identify those area of shoreline that may be particularly sensitive to oil spill damage and to help ensure that damage due to the use of inappropriate cleanup procedures is minimized. One frequently used vulnerability index was devised by using the assumption that coastal geomorphology frequently determines the types of ecological communities found there. Information about points of access to a spill site is essential in the event that men and equipment must be deployed at shore notice. Therefore, accessibility mapping is an important adjoint to vulnerability mapping. Coral reef vulnerability to oil spill damage will vary with reef type, zonation patterns, and tidal actions For example, in a fringing reef, the seward facing reef creast is generally exposed to high wave stress; here oil will probably have a short residence time before being dispersed naturally. In contrast, wave action is a lot less severe in the shallow reef flat that frequently supports seagress beds.

Here, a long oil residence time with a correspondingly high risk of oil damage is likely. Similarly, oil that strands on reefs with a limited tidal range will likely tend to persist for longer periods than that in areas with large tidal ranges. Although vulnerability and accessibility mapping techniques are now well established, little evidence suggests that such techniques are being applied to reef ecosystems. Some reef areas are well mapped and information is readily available on the location of refineries, tanker terminals, oil platforms, and major tanker routes. (e g., International Union for Conservation of Nature and Natural Resources 1980) This information can be used to produce maps that help locate areas of high potential risk. Newly developed techniques for mapping large areas of coastline, such as low altitude video overflights seem particularly suitable for use in reef areas but have yet to be applied.

Cleanup and Treatment of Spilled Oil

The sensitivity of reef ecosystems and the likelihood of their being impacted by ail will depend upon a variety of factors, including the quantity and type of oil and the degree of weathering. The likelihood of ecnomological impact will also very according to whether the reef is an emergent, shallow submergent, or a deep water type. The field and laboratory studies reviews indicate that the likelihood of damage is increased if the oil is incorporated into the water column of if it comes into direct contact with the coral surface. This, therefore, places the margent or shallow submergent section of a reef as areas of high potential sensitivity.

It also indicates that any cleanup of treatment attempt should endeavour not to enhance the transport of oil into the water column or into the permanently submerged sections of a reef. Careful consideration should be given to the treatment of oil in the vicinity of a reef as situation may develop where the cleanup or recovery attempt may be more damaging then the oil alone. Baker (1970) has pointed out that merely the shipping activity associated with the cleanup attempt may be damaging due to sediment resuspension by propellers and anchor drag breaking up corals. Although practical experiences of oil cleanup in the vicinity of coral reefs are lacking, mechanized techniques for containment and recovery of oil using booms and skimers are acceptable as they do not cause oil to sink or become incorporated into the water column.

Any restriction on use of these techniques in coral areas is likely to be brought about not by the risk of ecological damage but other environmental considerations such as high current speed, servere wave action, or shallow water depth that make their deployment impossible. At present, only few data are available on the impact of dispersants and dispersed oil on corals. Most information that is available has been derived from laboratory studies. Laboratory studies have also been carried out on biota associated with reefs. Both categories of study noted damage by dispersed oil, but the applicability of these in vitro data to a field situation is questionable.

Similarly, these studies used early generation dispersant which of themselves, where toxic to the carats. Using dispersants currently, available, indicated that concentrations of 20 to 50 ppm dispersed oil in a 24-hours exposure produced temporary stress reactions in corals. Few deleterious effects from 24-hour exposures at 1 to 5 ppm were observed. An oil removal technique that cannot be recommended for use in coral areas is the application of particulate sinking agents. These materials absorb oil and because they have a density higher than that of sea water, sink with the oil attached.

Deleterious effects could be expected from the sunken oil coming into direct contact with the corals. Several methods currently in the developmental stage appear to be potentially suitable for use in coral areas and other marine environments likely to be sensitive to oil. Among these are agents that have been developed that, when added to oil, gel in a semisolid from that can then be recovered. Research is also being carried out to develop nutrients that accelerate the bacterial biodegradation of oil. Several studies have shown that corals are seemingly unaffected by a layer of oil floating on the water surface. Therefore, where unsuitable hydrographic conditions or problem of access preclude the use of mechanical containment and recovery techniques, it is probably best, given our present knowledge, to leave the oil alone to weather naturally.

Measures of Damage at the Species Level

Structural changes occur on a reef when a species or combination of species that are dominant components of a system are replaced by other species are eliminated from the system. If the replacement of elimination or a species is significant, then community, or functional level changes can become evident. Changes in species richness, coral cover, and species diversity are potential criteria for the assessment of catastrophic structural damage to corals and any subsequent recover. Loya (1972) has modified the Shannon-Weaver diversity measure for use with corals as well as any encrusting assemblage.

Grigg and Maragos (1974) used these parameters to describe recolonization by hermatypes on lava flows, as did Sheppard (1980) when assessing the impact of harbor construction on corals. Porter *et al,* (1981) used species richness, coral cover, and relative abundance to asses structural recovery of hermatypic corals following hurricane damage. In a recent paper, Pearson (1981) argued that measures of diversity, percentage cover, and similarity taken individually do not provide useful measures of either damage as recovery. He suggested that measures of damage and recovery should totally incorporate measures of percentage cover, mean colony-height, surface-index, species diversity, and similarity indices. In addition, because reefs are dynamic systems that can undergo change due to numerous environmental factors, any observed structural changes should be evaluated in terms of the structural characteristics of neighbouring reef systems.

Measures or Damage at the Community-Level

The structural diversity of reef ecosystems is paralleled by their function a complexity. When attempting to assess damage and recovery of coral reefs in functional terms, to useful line of investigation may be development of indices that provide a measure of total reef functions or malfunction. One major functional characteristic of reef ecosystems is their high rate of primary production that may be 10 to 100 times greater than that of the phytoplankton in surrounding waters. Much of this reef primary production is attributable to benthic algae.

A principal determinant of this high rate of productivity in nutrient poor areas is the retention and recycling of nutrients within the reef. This feature may be useful means of determining the status of total reef function. For example, Pilson and Betzer (1973) food that, although concentrations of phosphates in water passing over a reef were low, there was no difference in the upstream and downstream levels, thereby indicating that a reef recycling mechanism was present. Evidence of phosphate recycling has been found by Pomeroy et al. (1974).

There is also evidence of ammonium and nitrate recycling mechanisms. An increase in nutrient loss suggestive of a breakdown in nutrient recycling mechanisms may be a potential indicator of a total reef malfunction. Further evidence for the steady state nature of a reef ecosystem is provided by the ratio of gross production to community respiration (P/R) that frequently approaches unity. Again, any large be change in P/R ration could, be indicative of reef malfunction.

Coral Reef Recovery

Our ability to measure recovery in a coral reef system binges on the success of a program to define or measure than is damage. Endean (1977) said that complete recovery "involves the return of those species typically associated with corals and forming part of the coral reef community in the particular geographical region concerned and the reestablishment of the complex relationships normally existing among the species," While this definition is theoretically correct, from a practical sense in terms of measuring recovery, it is probably too broad. Coral reefs are among the most complex marine ecosystems in terms of communities of structure and ecological function. Monitoring all aspects of reef biology in an attempt to confirm a recovery or to define a stage of recovery is not feasible.

There are, however, aspects of reef structure and function that could logically be used as indicators of a level of recovery. Specifically, a coral reef has patterns of species composition, abundance, dominance, reproduction, recruitment, growth, and mortality among the hormatypic organisms. A gauge of recovery may be generated by comparing qualitative observations and quantitative measurements of these pattern on the recovering reef with those presumed to be characteristic of the chosen model of a fully recovered reef at the site in question. This methodology, however, requires the assumption that recovery of the coral organisms will eventually result in the recovery of the entire reef community. Few reefs in the world have been subjected to quantitative ecological assessments. The means that it is unlikely that any site-specific information will be available for comparative purposes. However, reef morphology and biotic community structure are typically similar over a definable range of variation within the area or region.

Therefore, there is likely to be a basis upon which to roughly estimate the former structure and condition of nearby unimpacted reefs existing under similar conditions of natural environment, along with the remains of the impacted reef. If it is possible to infer pre-impacted structure from regional patterns, there is at least a structural and functional model for a recovered reef at the locality in question. The measurement of recovery is complicated by the fact that recovery on coral reefs tends to be a long term process Endean (1977) found that the recovery period is correlated with the extent of the damages inflicted on the bard coral cover. Small scale localized destruction from natural events generally requires less than 10 years for recovery provided that major sectors of the community remain virtually intact and the area is conductive to coral growth. Heavy destruction requires 10 to 20 years for full recovery, while especially severe impacts may require several decades for complete recovery. If a chronic source of pollution is present in the area of the damaged reef recovery may be further prolonged or may not occur at all.

Cleanup

What cleanup measures are appropriate and what are their relative effectiveness are critical questions regarding oil spills in tundra and taiga. Whenever possible, hydrocarbon spills should be avoided or their effects mitigated. Several measures can be employed to help accomplish this. All above ground fuel storage containers should be contained within a pit or depression lined with a hydrocarbons in previous layer. Oil pipelines should be designed to withstand weathering, corrosion, and limited displacements such as those due to earthquakes. Finally, whenever possible, oil pipelines should be routed to avoid sensitive ecological stress such as populations of endangered species. Unfortunately, containment or avoidance is not always possible, so cleanup measures for the various soil and vegetation types must be considered. Once a spill has occurred, choice of cleanup strategies will depend upon the type of oil spilled, the nature of the spill the amount of oil, the season of the spill, and the location of the spill. Site aspect, vegetation, and soil types are critically important factors.

In addition, the presence of permafrost in must areas of tundra and much of the taiga will preclude the use of many cleanup methods routinely used in warmer climates or non permafrost areas. For example, in permafrost areas with a high ice content any surface disturbance, such as removal of the oil-soaked organic soil surface horizon, tillage, or vehicle movement will induce thermal degradation of the permafrost and subsidence (thermokarst), which will increase rather than decrease total environmental damage of the site. Under such conditions, passive (or no active) cleanup should be considered.

Passive measures may also be the best means of facilitating natural recovery in cases of small spills of crude oil or contained or limited spills of industrial oils with high proportions of light fractions, much of which will be readily lost by volatilization. However, recent work by Linkins and Feteher (1983) has shown that residual oil in the Oe-OI horizon of tussock tundra soils may alter reproductive and biomass allocation patters of Betula nana and Eriphorium vaginatum such that their successful reestablishment in oil-contaminated soils is questionable. When active cleanup measures are necessary there is heavy reliance on site access for transporting equipment and manpower to the spill. In remove locations, this can involve considerable cost as well as be environmentally damaging unless it can be accomplished with the appropriate equipment, especially during the persnow thawed soil times of the year or snow preseason. Generally, the choice of which active cleanup techniques are used depends upon the time of year when cleanup most occur.

In winter periods when the ground is frozen or snow covered both access and cleanup are facilitated. The dormant stage of the plants and low winter temperatures that aid in cooling and congealing crude oil 110 that its spread is limited both contribute to minimizing damage. Snow also acts as an absorbent and cushion protecting surrounding vegetation from oil exposure as well as vehicle damage. Generally, the oil-contaminated snow can be mechanically scraped off the site removed, and placed in properly contained areas such as was done at the February 1978 Steele Creek spill along TAPS.

Burning the spilled hydrocarbon may also be used as a cleanup measure either before or after mechanical removal. Combustion may be difficult to initiate and generally will be incomplete. However, burning can help to remove the more toxic light fractions of the oil if done soon after the spill occurs. Small spills may be difficult to ignite if not immediately burned, but large spills seem to burn readily even several months later. This may be due to a large proportion of the light fractions being rapped within the soil in large spills as surface weathering produces as extensive asphaltic cover. Burning is relatively inexpensive, may leave a tarry residue on the surface that can inhibit microbial decomposition of the oil and vegetative recovery. Finally, following any winter cleanup procedure, appropriate measures should be taken to ensure against, remobilization of the oil during snow melt.

Such measures as absorbent booms, straw, or in non permafrost and thaw stable permafrost areas, containment ditches or dikes may be used. The hydrocarbon will then be intercepted or contained to facilitate removal. The choice of active cleanup measures during the summer or snowfree period will depend upon the phenological stage of the plant, the soil moisture content, depth of the soil water table, and ice content of the permafrost as previously discussed.

Mechanical removal of oil contaminated soil should be considered only at times when potential damage to plants is great from prolonged extensive contact with the hydrocarbon, the ice content of the permafrost if low, and there is relatively easy access to the spill site. Also, the volume of the spill should be such that natural recovery is unlikely and there is high potential for extensive contamination of adjacent areas through contained movement or later remobilization. If removal of the contaminated soil is not possible, burial can be considered. However, burial should be done only in thermally stable, frozen material since subterranean remobilization can occur and cause contamination of adjacent areas.

Oil may also be removed by floating the oil and vacuming it off the surface of the water. This was used successfully at the TAPS value 7 spill in July 1977, ready access to the site as well as availability of a large volume of water, a wet site, and feasibility of containment. However, in some cases, oil may continue to appear on the surface of the water for a year of more after toe spill as it is displaced from the surface organic horizon. Finally, manual cleanup with minimal soil disturbance should be considered in permafrost areas. Absorbents such as straw or commercial materials can remove much of the oil present on the ground surface. Although costly and very time-consuming, this method will minimize any physical disturbance and decrease the chance of thermal degradation of the spill and adjacent sites. The extensive reliance on reseeding with non-native species and long-term fertilization, whereas a relatively easy but expensive means of regeneration, does not guarantee optimal long-term regeneration or restoration.

Observations of natural restoration on a 1949 disturbance and on a 1970 disturbance at Barrow, Alaska suggest that it may be preferable to let some hydrocarbon disturbances recover naturally through reinvasion by native plants in lies of reseeding with exotic plants. Recent work discussed by Shaver et al. (1983) that shows that the shallow organic soil horizons can serve as a major seed bank for regeneration suggests that preservation of the upper organic soil horizons may also be preferable if possible to maximize potential for native plant reestablishment. Their discussion also points out the possible deterrents of fertilization on delaying restoration of the natural vegetation. Penetration of the hydrocarbons into soil plant rooting zone as previously discussed wilt influence the choice of regeneration/ restoration techniques.

On wet sites where there is limited penetration of the hydrocarbon into the soil rooting zone, natural recovery by the native vegetation may be sufficient if cleanup activities adequately remove the surface hydrocarbon contamination. On drier site, however where the hydrocarbon will rapidly penetrate the soil rooting zones, killing the majority of the vegetation, it may be necessary to institute extensive soil cleanup, rapid revegetation efforts. This will be especially true if site stability (othermokarst, slope erosion, etc.) is a problem.

Regardless of the cleanup, revegetation/restoration activities it should be reiterated that care should be taken to preserve of site integrity. Slower spill site revegetation/restoration should always be preferable to risking both on and off site degradation. Hydrocarbons spills in the Arctic and subarctic cause both short-and long-term effects on the soils microbial populations, vegetation, and wildlife. Although a number of descriptive studies on the effects of hydrocarbon spills have been conducted, these have generally focused on short-term effects.

Furthermore, there has been very limited development in our understanding of the important mechanisms of the direct toxic effects of oil and the effects of altered soil characteristics on organisms. When our ignorance in these areas is compounded by our limited knowledge of the in site rates and nature of hydrocarbon degradation in cold soils, it makes it very difficult to estimate effectively the duration of the influence of hydrocarbon in the soil. Likewise, it is very difficult to predict which cleanup, revegetation techniques will be most successful in alleviating hydrocarbons related deterrents to microbial population and vegetation growth. Development of better methods for restoring Arctic and subarctic hydrocarbon spills requires a commitment to long-term research efforts focusing on integrated cleanup, revegetation/restoration research.

Research should focus on the development of integrated postspill functionally based ecosystem efforts to determine and evaluate the relative importance of factors limiting site restoration. These studies should be connected on new sites that have been thoroughly described and on existing spills sites where adequate documentation exists as to the specific cleanup and revegetation activities that were employed. Efforts should also focus on natural hydrocarbon seep areas and long-term unaided spill sites where native vegetation exists. Integrated research efforts directed toward these goals as well as toward associated off site cleanup and revegetation disturbances should then begin to provide the information necessary to formulate effective revegetation/ restoration programs in Arctic and subarctic areas. Unfortunately, until research is initiated, there will be limited advancement in more effective, less costly revegetation/restoration practices.

12

Pollution Abatement

Biotechnology in Reduction of CO_2 Emission

Increase in concentration of CO_2 in atmosphere has of late become a global issue and is linked with rise in atmospheric temperature causing the 'Green House Effect'. GH gases include CO_2, CH_4, CFC and water vapour. Of these gases $\cdot CO_2$ is agent and is believed that during the last 150 years, its level has increased by 25% leading to increase of 0.5°C in Northern hemisphere and global warming in general.

Present models suggest that for doubling of CO_2 level in atmosphere, global temperature can increase by around 1.5–5°C and would lead to rise of sea levels, odd rain-fall, shifting of vegetation zones and their effects on agriculture in general.

This led to the devising means of reduction of CO_2 emissions. Along with chemical approaches, the possibility of biotechnological reduction is also being thought and investigated in different parts of world as an alternate approach.

Photosynthesis as a Means of Reducing CO_2 Emissions

Photosynthesis by plants is considered to be an obvious means of energy efficient biotechnological reduction of CO_2 released from industrial exhaust, and is summarized as:

$$nCO_2 + 2nH_2O \xrightarrow{\text{Sunlight}} 6(CH_2O) + nO_2 + nH_2O$$

In this respect the photosynthetic efficiency of both terrestrial and aquatic vegetation can be increased.

Higher plant photosynthesis

Violent deforestation has resulted in significant reduction of tropical rain forests are getting diminished every year. Their conservation is an urgent need and for that matter, 2 approaches may be made. On the one hand, selection of fast-growing trees and their propagation is done and on the other, recourse to biotechnological methods like 'micro-propagation' and 'synthetic seed' production through tissue cultures are attempted at different centres.

Another biotech, approach for genetic improvement of photosynthetic CO_2 fixation is also desirable and is known that the enzyme RUBP-case is one responsible for CO_2 fixation by plants. Its two sub-units are controlled by nuclear and cytoplasmic genes. Attempts are being made to genetically manipulate this enzyme in crop plants to increase the photosynthetic efficiency. Some approach could be made in forest trees at the tissue culture level before their propagation, so that regenerated plants from culture would have more CO_2 fixation potential.

Microalgal photosynthesis

This is an efficient mechanism for CO_2 reduction, if well designed *photobioreactor* could be constructed for their intensive cultivation. Due to high rate of photosynthesis, they appear to be good candidates in this respect. The *photosynthetic quotient*, which is a ratio of release of O_2 to CO_2 fixed, is usually >1 for microalgal. This means that during process, they can generate more of O_2 than amount of CO_2 consumed per unit biomass than in case of higher plants. Moreover microalgae offer economics of scale, as they can be grown by volume instead of area, so photobioreactors would be useful where cultivation be made in cultures with artificial light and fibre optic light radiator. Several such reactor design have been proposed and large scale *Chlorella pyrenoidosa* and *Spirulina maxima*.

It has been estimated that amount of CO_2 released from a power plant of 150 MW capacity would require an algal culture volume of 15×10^4 m^3. Moreover, such a uge microalgal biomass could be effectively used for extraction of protein as food and feed, and is known that both *Chlorella* and *Spirulina* contain high amount of protein per unit biomass among the plant kingdom.

Industrially generated CO_2 is much higher (20%) than its normal level in air. CO_2 reduction from such high concentration would thus require algal forms or strains which can tolerate high CO_2 level, eg., increased alcohol tolerance by yeasts. In this respect, genetic manipulation with chemical mutagens like EMS or NMU and other yielding higher CO_2 tolerant mutants in *Anacystis Nidulans* and *Synechococus* in encouraging. This has also led to discovery of a number of strains, with higher CO_2 tolerance, of other algae like *Chlorococcum* and *Oocystis*.

Alternatively, using gene cloning technique may perhaps be possible to develop strains with faster growth rate to yield larger biomass of concerned algae. Changes in sedimentation and flocculation behaviour would be another advantage with respect to harvesting algal biomass as it can be a part of overall process economics.

Reducing CO_2 Emission from Sea Water through Biological Clacification

Deep sea offers viable reservoir for long-term storage of CO_2. This is achieved with help of clacifying organisms like corals and a number of colcareous green and red algae. They live in a symbiotic relationship and the clacification process is rapid.

$$H_2O + CO_2\ (aq) \rightarrow H_2CO_3$$

$$H_2CO_3 + Ca^2 \rightarrow CaCO_3 + H_2O + CO_2$$

If algal species can be found out which would tolerate high pH and carbonate, then photosynthesis by such algae may lead to precipitation of $CaCO_3$.

Algal Photosynthesis in Waste Water Treatment

Waste water from human work, are of municipal, agricultural and industrial origin. Due to diversity of sources of these wastes, their features also vary extremely. They contain organic and inorganic sub stances, pathogenic organisms and various toxic materials like heavy metals. So treatment of wastes of purification is rather complex.

The treatment processes may be categorized as physical and biological. The physical process consists of straining and settling. This cannot remove a major portion of organic matter, which remain as colloidal and as dissolved solids. This can be achieved by what is called phenomenon of *mineralization* of organic matter, i.e., breaking down into simpler forms through bacterial oxidation or *biodegradation*. This process need steady supply of oxygen as respiratory requirements of bacterial flora in waste. In conventional biological practices, the oxygen is supplied artificially by continuous 'Bubbling' of atmospheric air. Aerobic *oxidation ponds* or *stabilization ponds* are used throughout the world for such treatment. But it needs costly equipments and manpower.

It was long known that micro-algae can serve rich source of oxygen in fish ponds. This was first used in sewage treatment by Hermann and Gloyna in 1958. The algal system can supply the oxygen needed

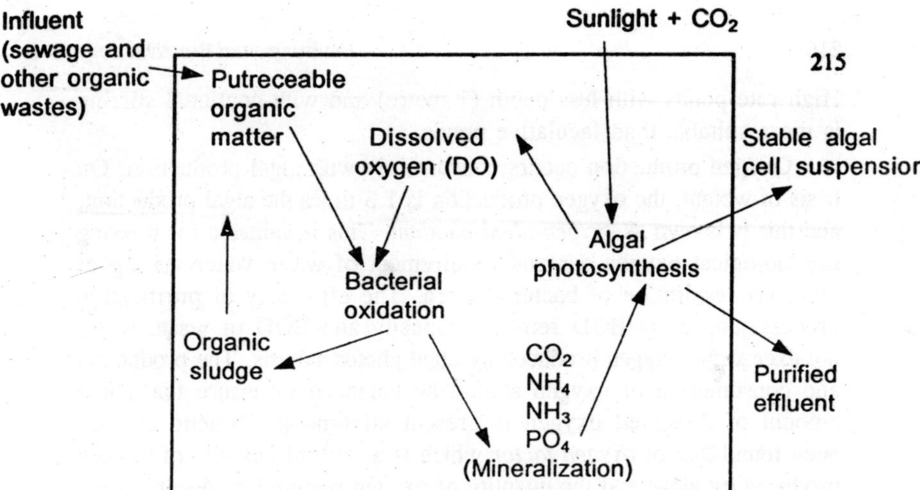

Fig. 12.1. Algal-bacterial symbiosis in photosynthetic oxygenation.

by bacteria and protozoa in process of degradation of organic contents of waste water and is a natural process and therefore no artificial aeration is necessary. The conventional treatment of sewage and other waste waters consists of three phases, viz., primary, secondary and tertiary. The algal treatment is introduced at secondary stage where liquid effluent from primary stage is taken care of and bio-oxidation takes place.

Oxygen Production by Algae

The process of *photosynthesis* is known as oxygen generator, where Co_2 and H_2o are used by plants to build carbohydrates and release O_2 in presence of sunlight. The photosynthetic efficiency of many micro-algae is very high. Under satisfactory environmental conditions with availability of enough sunlight, as in many tropical countries, the oxygen produced by these algae surpasses their own respiratory requirement. This surplus oxygen be made available in oxidation pond by the culture of specific algae. It is an inexpensive and natural process.

The species of different algal genera which are best suited for purpose are mostly members of green algae or *Chlorophyceae*. Some of them are *Chlorella*, *Scenedesmus*, *Euglena*, *Chlamydomonas*, *Hydrodictyon*, *Ulothrix* and *Tribonema*. They have high 'oxygen donating' power and are very useful in the biotreatment of waste water. This depends on steady flow of CO_2 in pond water. It has been estimated that 10-20 ppm of CO_2 is optimum for efficient O_2 donation. Again, in carbon limited system, the pH may get increased upto 10, while the optimum pH for bacterial growth and action is about 8.3. So a balance is necessary for proper symbiotic function of algae bacteria system.

High rate ponds with less depth (1 metre) and with continual stirring is more suitable than faculative ponds.

Oxygen production occurs concurrently with algal production. On basis of weight, the oxygen production is 1.3 times the algal production, and this is known as *oxygen-algal quotient*. This is valuable for meeting the biological oxygen demand requirement of waste water arising of vigorous respiration of bacterial flora. The efficiency of purification process judged by BOD removal capacity and BOD of waste water not exceed the oxygen produced by algal photosynthesis. The production and consumption of oxygen should be balanced to ensure that some amount of dissolved oxygen is present all time in effluent and has been found that of *oxygen factor* which is a ratio of quantity of oxygen produced by algae and the quantity of oxygen required by decomposing bacteria, is maintained at 1.5 to 1.6, high level of BOD removal can occur. Result show 80-90% BOD removal can be achieved in course of 30 days.

Moreover through this algal treatment, it is possible to precipitate some of toxic substances like heavy metals from waste water. This is due to the high negative charge on algal cell-surface against strong positive charges of heavy metal cations which absorbed and settle out.

One of problems, associated with use of algae in liquid waste treatment is separation of algal biomass after they have done their job. A portion of algal cells may die during long retention time and may get settled at bottom along with other particulates and thus create anoxic situation. This may be prevented with use of surfactants and finally, can be separated out by mechanical strainer.

The conditions is to be satisfied for proper algal treatment of liquid wastes are the presence of bright sunlight helping higher photosynthesis efficiency vis-a-vis higher oxygen production, a temperature below 35^{0}C to prevent thermal inactivation of algal cells, steady supply of carbon source to prevent pH increase and continuous slow stirring to prevent development of anoxic condition.

Apart from purification of waste water, such pond could be a source of valuable by-products. Algal cells rich in protein could serve as food and feed or be used as measure. Being rich in DO, such ponds may well support fish life, and algal cells serve as fish food and it also be a means of recovering metals present in waste, as contaminant by way of algal cell surface adsorption.

Thus whole process of algal treatment of waste water is useful in many ways, apart from primary objective of purification and possible reuse of such water. This system also reduces the risk of large number of pathogenic microbes getting into effluent as many of them either become dead and get settled at bottom of pond or are screened out along with removal of a algal biomass.

Eutrophication, Algal Blooms and Biological Phosphorus Removal

Eutrophication

The term 'Eutrophication' is derived from Greek work meaning 'rich', while eutrophic water refers to stage at which it is enriched with undesirable organic or inorganic nutrients containing phosphorus and to a lesser extent nitrogen. Natural sources are nitrogen and phosphorus cycles and man made sources are agricultural and municipal wastes containing fertilizers and detergents as casual agents. Undesirable amount of these elements in waste water favour excessive algal growth and lead to *eutrophication*. It causes depletion of oxygen, foul smell through generation of sulphides and death of non-resistant organisms. It results in the colonization of such water by a particular group of organism, cause sedimentation and eventually gets filled with the resistant types. Such a situation has been termed as 'water bloom'. Dissolved organic compounds act as direct source of nutrient for huge algal growth or through bacterial activities release more CO_2 favouring algal photosynthesis.

Among algal species, members of microalgae, eg., Cyanophyceae are predominant. Different species of genera like *Lyngbia*, *Microcystis*, *Anabaena*, *Aphanizomeum* and *Oscillatoria* are involved. Atleast 20 algal species are capable of forming 'blooms'. These have gas vacuoles in their cells which help them to float and survive for varying periods of time. In a dense bloom, gradual oxygen deficiency results in collapse. This leads to further release of nitrogen and phosphorus and helps in development of another bloom. It has been termed as *bloom succession*. In a mature bloom with excessive growth of cyanophycean alga, *Microcystis aeruginosa*, a reddish scum is produced. Seeing this from a distance, one can predict that the lake or pond has undesirable amount of phosphorus. Thus it serves as an *indicator* species. In coastal areas the so called 'brown tide' is caused by a golden brown alga called *Aureococcus*.

Algae absorb phosphorus and store it as polyphosphates within the cells. Phosphorus concentration above 6 ppm favour explosive growth. It has become a global problem as ponds and lakes gradually turn into marshes, which is why in many western countries, phosphorus based detergent manufacture is discouraged. Algae are considered to be biological means of N_2 and P_3 removal, which also help checking eutrophication.

Apart from Cyanophyceae, dinoflagellates like *Ceratium* and *Cryptomonas* also present in blooms. In marine conditions, dinoflagellates cause red or brown colouration of sea. Centric diatoms and seaweeds like *Ulva*, *Enteromorpha*, *Cladophora*, *Gelidium*, etc., have been used for assessing eutrophication of water in marine environment. Models have been proposed for prediction of eutrophication. Literature on eutrophication have been reviewed by Porcella (1978).

Such heavy growth of *bloom* algae release endotoxin like LPPS affecting fishes and birds and causes some diseases among human beings, namely gastroenteritis, diarrhoea, nausea, etc.

Control of algae in eutrophic waters be done chemically or biologically. Chemical means constitute use of algicides like copper sulphate, sodium arsenate, 2,4-D and 2,3-DNQ. But all these increase the sludge volume from sedimentation. They also be controlled with use of *cyanophages*. This has been found to be effective by a number of such viruses, viz., LPP-1, A-1, N-1, AS-1, etc. Although these viruses, being specific for BGA and do not pose problem to other aquatic organisms yet the dead algal cells may release toxins in water.

Biological Phosphorus Removal from Waste Water

As phosphorous is ecologically significant in algal productivity, its removal from aquatic bodies is essential to protect them from eutrophication, usual practice is to precipitate them chemically with salts of Ca, Fe, Al and Mg. With calcium salts, phosphorous is precipitated as hydroxypatite, $Ca_5OH\ (PO_4)_3$.

$$5Ca^{2+} + 7OH^- + 3H_2PO^-_4 \rightarrow Ca_5OH(PO_4)_3 + 6H_2O$$

However, the alternate method is biological one, where phosphate metabolizing bacteria help in process. The energy required for this is made available by release of phosphorus bound as *polyphosphate* in volutin granules, in the bacterial protoplasm.

The bio-P removal has advantage that additions of chemicals can be avoided and helps the reduction of sludge volumes. The principles of biological treatment plant lies in exposure of organism to alternating anaerobic and aerobic conditions. This can be achieved with or without nitrogen removal. When nitrogen removal is needed, an anoxic condition be introduced in between.

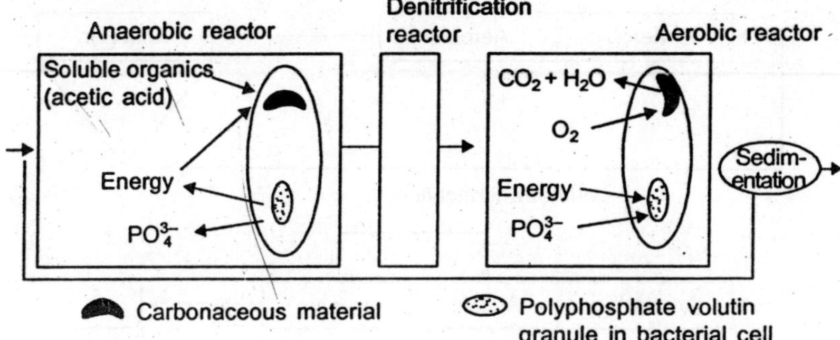

Fig. 12.2. Biological release and uptake of P in alternating anaerobic and aerobic condition.

Under anaerobic conditions, transport and storage of simple organics such as acetates require energy which is got from polyphosphate reserves of bacteria with the release of phosphates. While under subsequent aerobic stage, the organic matter is oxidized to produce energy and reaccumulation of phosphates into polyphosphates. The net effect in the excess of phosphorus in bacterial cell.

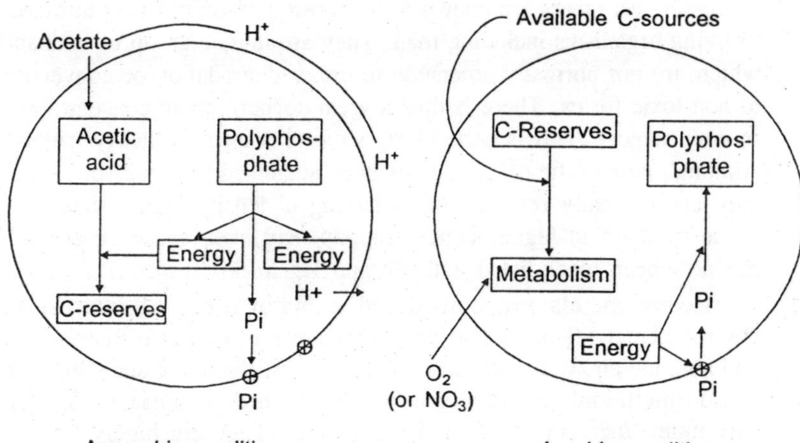

Fig. 12.3. Models for P removal in bacteria through anaerobic and aerobic metabolisms.

The possibility of removal of nitrates from waste water with use of immobilized cells of the bacterium *Pseudomonas fluorescence* is a continuous process shown by Tenegerdy et al., (1981). A combination of bio-P removal with simultaneous chemical precipitation is likely to achieve low effluent phosphorous concentrations.

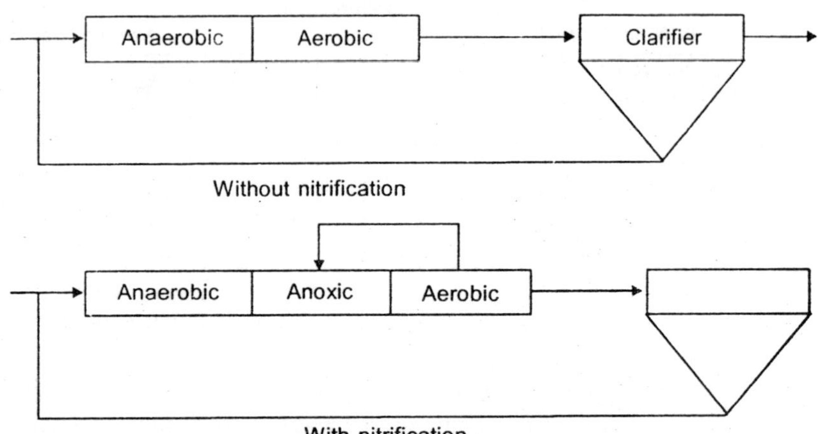

Metal Pollution and its Bio-abatement

Metals and metaloids are everywhere in the biosphere. Rapid industrial development on other hand is gradually redistributing them. Through some metallic ions are essential for continuation of life processes, yet others are known to be harmful at various concentrations to living organisms including man. They are unique group of toxicants which are not normally amenable to either degradation or conversion to non-toxic forms. There is thus a great concern about concentrations in environment. This is particularly so with respect to 'heavy metals'. Although no specific definition has been attributed to groups of metals, they are normally referred to as having a density higher than five. Some of them in higher concentrations may even cause cancer. Cd has now been black-listed with Pb and Hg in this respect.

Heavy metals are considered a major field of interest by environmental scientists and engineers alike as they are increasingly polluting the air, water and soil. Some of most important and dangerous metals/mettaloids in this group are arsenic, cadmium, copper, chromium, mercury, lead, nickel and zinc. They are highly toxic to living systems and some cause cancer in long run even when injected in very low concentrations. Toxicity is not dependent on concentration and exposure, but is also known to increase with electropositivity of metals, viz., Hg > Cd > Zn. Exhaustive studies throughout the world have not only yielded knowledge of physio-chemical properties of these toxicants, but also their biological impacts. Different solubility, mode of transport, interactions among different metal species, adsorption or absorption, chemical complexation with organic ligands, non-

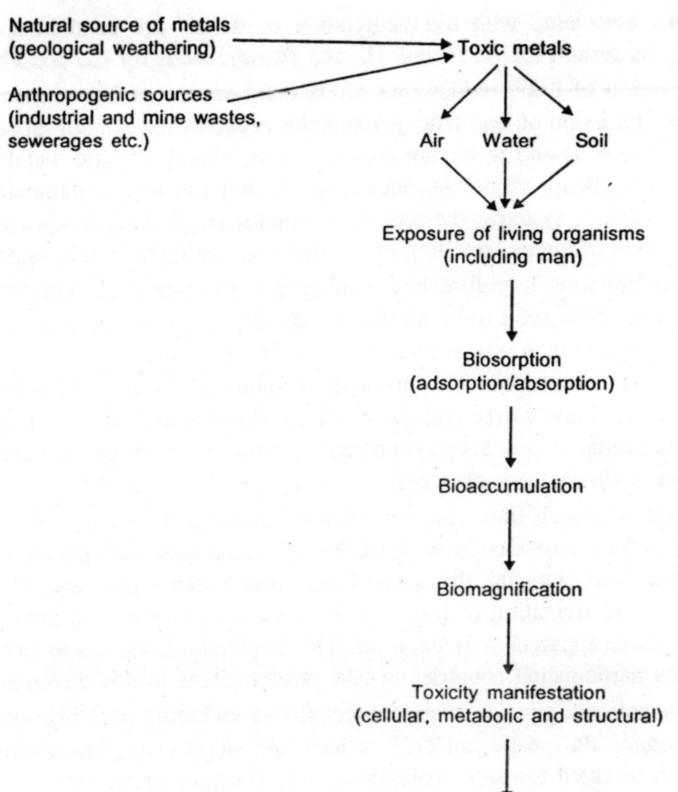

Fig. 12.4. Source and fate of toxic metals in biosystems.

degradability, persistence in environment, bio-accumulation, bio-magnification in food chain and biomethylation are some features which are considered.

Biomagnification is phenomenon where the metal, absorbed by organisms at lower trophic level of food chain, gradually get more concentrated in those organisms which are higher levels where man is the ultimate victim (At the higher trophic level of food chain man is there—so gradually concentrated metal ions enter human body—causing maximum damage. *Bio-methylation*, on other hand, involves the transfer of methyl group (CH_3) from organic compounds to metals die to microbial activities in soil and water, through this some metallic ions get detoxified, yet others turn into more dangerous forms due to methyl conjugation. Methylated arsenic is less toxic than As (III or V). But methyl mercury (CH_3—Hg) is more toxic than the metallic ion itself. Some of microbes associated with bio-methylation of metals are candida and methano bacterium for As *E. coli* Hg and *Pseudomonas* for Cd and Pb.

Emission of toxic oxides into air is main source of heavy metal pollution. Emission of lead from automobiles accounts for 2/3rd of global input, while soils and water are considered as 'sinks' of toxic metals. In soils, they being immobile, accumulate in top soil and contaminate crops. In aquatic systems, the sources are industrial discharges sewage, metal mining effluents, smelling etc., other sources include acid rains.

A voluminous literature have gathered with respect to pollution menace and these need to be addressed suitably to tackle the problem effectively not only at source level but also through abatement measures to minimize the impact. The problem is rather acute in developing countries like India where resurgence of industrial activities in making a quantum jump in process of economic upliftment and threat of metal pollution is slowly becoming real.

There had been little concern for their presence in waste waters, so long as they remained at subtoxic levels. Since classical episode of 'Minamata' and 'Itai-Itai' disease in Japan, and mercury poisoning due to wheat field irrigation in Iraq, the seriousness of problem on human health became apparent at global level. The Stockholm Congress in 1972 urged the participating countries to take positive steps in this direction.

Conventional physico-chemical techniques including precipitation, ionexchange, etc., were not only refined and stepped up, but efforts were also directed towards evolving suitable alternate biotechnological protocol. Biosystem approach has not only helped in better understanding of mechanism of metal toxicity and detoxification, and device effective mitigating measures, but also be opened the possibility of using these bioimpacts as pollution monitoring parameters. Many of aquatic plants, including photoplanktons and benthics are gradually show their unique sorption potential of metals from bathing media and act as *natural bioscavengers* of effluents. Many microbes also have metal detoxification potential. Biosurveillance of toxic metals is still in informative stage and is expected to turn out to be a real cost effective device in long run. This is inexpensive and effective too. This approach would also help in recovery of many industrially valued metals from further treatments of used biomass alongside the generation of biogas as energy source. The recovery of metals, however, is facilitated when dead or immobilized biomass is used as a sorbant in adsorption process rather than absorption.

Environmental biotechnology has now added a new dimension to pollution monitoring and abatement, while different environmental protection agencies the world over are appreciating the role of the technology, particularly for small scale industries in developing countries, enhanced research effort in this area is a result of renewed global interest in biotechnology. The use of different groups of plants including common aquatic weeds, different microbial species and various agricultural wastes as low-cost bioabsorbents of heavy metals have shown promise, and huge amount of data have accumulated in this regard.

Plants in Aquatic Metal Pollution Abatement

Higher plants (aquatic macrophytes)

These are observed floating weeds in ponds, ditches or streams and are mostly found in tropical parts of the world.

Since last 2 decades, these groups of plants have been widely tried for exploration of their uptake and accumulation potential of toxic wastes, particularly heavy metals. Attempts are being made on a variety of such plants in different centres with a view to screen out one or more such 'Natural Scavengers' of toxic metals. Some of the commonly used ones are discussed here.

Water hyacinth (Eichornia crassipes)

This is most common obnoxious weed found in tropical waters and is a problem to scientists engaged in devising methods for its control. However, its capacity to absorb Cadmium (cd) from the bathing medium at rate of 9.1 μg/1 and to help build up metal concentration at tip of plant at 6.1 μg/g of dry matter within 24 hours was first shown by Wolverton and Mc Donald, (1979), Chigbo et al., (1982) on other hand has shown the remarkable capacity of plant to absorb relatively high concentrations of metals like As, Cd, Pb and Hg over a short period of time. They noted that from a 10 ppm solution of each metal, the ratio of concentrations of As and Hg in the leaves to that in stems was 2:1, while incase of Cd, and Pb, it was about 1:1, O'Kneeffe et al. (1984) noted that Cd uptake by this plant was a biphasic phenomenon having initial rapid absorption phase of 4 hours and later at a slower pace upto 72 hours. By this time, it could remove more than 80 per cent of metallic ions from this medium. It has been noted by different workers that the plant has sorption potential of a huge array of metals without itself getting much affected. Pinto et al. (1987) suggested that water hyacinth be suitably used for recovery of valued metals like silver from mine wastes.

Duckweeds

Like water hyacinth, members of duckweed family have shown high promise in heavy metal sequestration. Tanaka et al., 1982 noted copper absorption by *Lemma* spp., while Nasu et al., (1984) observed differential mode of uptake response from a bimetallic medium of *Cadmium* and copper by *Lemna paucicostata*. From their findings on chromium removal capacity from waste waters by 3 duckweed species, viz., *Spirodela punctata*, *Spolyrrhiza* and *Lemna gibba*. Tripathi and Chandra (1991) confirmed the use of *Spirodela Polyrrhiz* in chromium removal. Chatterji and Nag (1991), on the other hand, shown how Lemna minor can be effectively used for removal of Cadmium, mercury and copper. Chatterji and his group noted that *Spirodela polyrrhiza* could effectively remove Pb and As from aquatic medium.

Though these plants are very small in size, they appear to have a remarkable in-built resistance capacity against metals. Hyacinth ponds and duckweed ponds are in use in many countries for waste water treatments containing metallic effluents. In India it has started only a few years back.

Water lettuce (Pistia stratiotes)

This is another free-floating aquatic weed common in tropical waters which has attracted attention in pollution studies. A number of authors have been shown good promise as metal Scavenger. Within a certain concentration range of metals, it can remove about 74 percent of Pb, 80 percent of Cd and 90% of Hg without showing any symptom of injury. Recently noted capacity of arsenic removal by Pistia upto 82% in 6 days under lab condition. According to present author, this plant is better candidate than others as metal scavenger.

Cat-tails (Typha species)

These plants usually grow in marshy areas with long erect strap shaped leaves. Taylor and Crowder (1984) used *T. latifolia* for removal of copper and nickel, while Oertzen and Finalayson (1984) used. *T. domengensistor* treating saline waters containing calcium and magnesium salts.

Apart from use of above mentioned plants with proven metal scavenging potentials, quite a few other aquatic also been experimented with and have indicated their bioaccumulation capacity of toxic metals, viz., *Hydrilla*, *Vallisneria* and *Potamogeton*. While water spinach (*lpomea aquatica*), also considered a poorman's vegetable, has shown its distinct capacity to absorb lead and chromium from waste water signalling a warning to the consumer of plant. It appears therefore, that so-called menacing aquatic weeds can have their beneficial uses too. There could be more unexplored. Not only can be used for such detoxification of water, many of these metals can be recovered subsequently by proper and treatment of slurry after biogas collection from huge biomass.

Also a number of land plants viz., *Ocimum* have been identified to decontaminate metal-polluted soil at mining and land fill sites. Lately Rufus Chaney and his group in U.S. (1995) have suggested green remediation of such metal affected soils by the hyperaccumulator plant species like *Thlaspi rotundifolium* (ray weed) and *T. coerulescens* (alpine pennycress), which could remove a number of metals like Pb, Cd, Zn, Ni, U and Co. The tiny plant *Arabidiopsis thaliana* is also a hyperaccumulator as noted by X, Yang of China and genetic control of such process is being worked out by use of mutants.

Lower plants and microbes

Besides the 'higher group' of plants discussed earlier, a good number of other comparatively simpler plants including microscopic organisms been tried for use as agents of bio-surveillance of aquatic metal pollution.

Ferns

The role of 2 fern genera Salvinia and Azolla have already been established in this regard.

The accumulation pattern of Cu, Cd and U in *Azolla filicoides* has shown it to be more in roots than in shoots. While 2 aquatic species of *Salvinia*, viz., *S. natans* and *S. rotundifolia* have demonstrated their ability of uptake of a large number of metals including Cr. No and Pb. The latter species been found to remove 95% Pb from a battery factory effluent in 15 days of batch culture. While S. Natans has recently been seen to accumulate As from aquatic medium.

Aquatic mosses and liver worts

Number of these plants, which rather uncommon, been found to occur in streams and rivers adjucent to Pb-Zn or Zn-Cr mining areas in different part of Europe & bio-concentrate high amount of metals.

Microbes

Algae

They may be constituents of phytoplankton, however, information is available on the bio-accumulation capacity of metals by member of both groups.

Trollope & Evans (1976) noted concentration of metals in fresh water 'Algal blooms', different species of single-celled green algal genus *Chlorella vulgaris* and *C. regularis* been noted to accumulate metals like Cu, Hg and U, while *C. pyrendoidosa* been used in recovering metals from sewage sludge. The net like alga *Hydrodictyon recticulatum* found to tolerate and accumulate high amount of Cu, Mn, Fe and Pb. Another green algal genus *Scendesmus* accumulates Ni and Cu Mishra (1985). Correlated Hg removal from chlor-alkali wastes to various combinations of blue-green algae.

Among the benthic macroalgae, members of genera like *Laminaria*, *Fucus*, *Ulva*, *Ascophylum* and *Codium* have all shown great minimize the metal pollution load of rivers. The nature of bonding between algal cell walls and metallic ions was noted by Christ, 1981.

Fungi

Selective accumulation of metals by different fungal species belonging to *Aspergillus*, *Pencillium*, *Rhizopus*, *Mucor* and *Neurospora* was noted by Nakajima and Sakaguchi (1986). The use of mushrooms like *Pleurotus*, *Agarcius* and *Volvariella* was pointed by Purkayastha and Mistra (1992). Tsezos and Volseky (1982) shown the mechanism of *Uranium* and *Thorium* biosorption by *Rhizopus arrhizus*. R.N.Kar and his group at RRL Bhubaneshwar are trying to use *R. arrhizus* biomass for sorption and recovery of copper from mine effluents of Hindusthan Copper at Ghatsila, Bihar. Mercury adsorption by yeast like *Saccharomyces* and *Sporobolomyces* was also reported by Strandberg et al., (1981). Tsezos (1984) showed that using Sodium-biocarbonate as eluent, uranium absorbed by *Rhizopus arrhizus* can be recovered and biomass can be reused many times.

Bacteria

Cell wall composition of bacteria plays a major role in this. Obaservations of Beveridge and Murray (1980) shown the site of metal deposition in bacterial wall. Nakajima and Sakaguchi (1986) also reported the use of various bacterial groups of actinomycetes in selective accumulation of many metallic ions. Similarly, *Bacillus circulans* and *Desulfovibrio sp.* has been noted to accumulate a number of metallic ions including copper. *Rhodospirillium* species was found to absorb a considerable amount of Cd, Hg, Pb and Ni from metallic solutions. Absorption of heavy metals like Hg by *E. coli* KP245, Cr by *Pseudomonas* species and U by *P. aeruginosa* are also on record. Strain improvement of microbial species through modern technology of microbial species through modern technology of genetic engineering is possible to increase the efficacy. Tengerdy 1981 and Tobin (1984) were of view that microbial biomass is a good sorbant of metals, while immobilized biomass could help metal reocvery.

Biomechanism of Metal Chelation & Detoxification

The uptake of zenobiotic metals be either adsorption or absorption and is possible through some physiological adaptiveness and homeostasis. The phenomenon is a complex one and may be categorized into the following.

Metal binding on cell wall

Cell wall is first barriers to encounter metals in plant cells. During passage of metal towards cell membrane, an amount of cations get bound with cell walls at the cation-exchange sites. Plant species differ in their cation exchange capacity with heavy metals in particular. Though metal binding with cell walls is rather common in lower group of plants, particularly fungi, bacteria and algae. Higher plants too have reported to 'fix' such metals in cell wall components and help detoxification. Different cell wall ligands, are basically bio-organic polymers, hold metals on cell surfaces till the binding sites get saturated. This has been shown incase of fungi.

Compartmentalization

This is also one of mechanism in plants to combat metal toxicity, plants by virtue of their properties transport the metals to various celllular compartments, especially to apparent free spaces, intercellular or intracellular, vacuoles and other such bodies like lysozomes to sequester the metals. This has been proved typically incase of zinc tolerance, where zincmalate pathway has been shown to transport zinc towards vacuoles. Zinc in intravacuolar bodies had been reported in root meristematic cells of grass species *Festuca rubra* L by Davies (1991). Ernst (1976) has shown zinc accumulation in lacitifers also.

Formation of immobilized heavy metals containing crystals have been described for Cd, Co, Fe, Pb, Sr and Zn and an increase of calcium oxalate crystals under heavy metal influence. In algae, the incorporation of metals into polyphosphate bodies as a tool for metal detoxification has been shown. These suggests that compartmentalization in an other homeostatic means of metal detoxification in plants.

Synthesis and binding with cellular proteins/peptides/phytochelatins

The presence of Cd-binding component in a fresh water blue green algal, *Anacystis nidulans* was first reported by MC Lean (1972). Later, however, metal binding proteins have been reported from all groups of algae. Fungi have also been shown to synthesize specific metal binding proteins/ peptides. An interesting feature of metal-ion homeostasis in plants been reported to be formation of a buffering molecule, 'phytochelatin' which is said to be present everywhere in all groups of plants. Fujita (1985) has isolated some metal complexing molecules from *Eichornia crassipes* upon treatment with Cu.

Grill (1985) isolated metallothionein like protein from *Rauwolfia serpentina* and confirmed it to be distinct from other peptides. This was known as 'Phytochelatin' for metal chelating properties in plants as opposed to metallothionein in animals. Various details have come up regarding this metal specific molecule mainly based on in vitro plant cell studies (Gupta & Goldsbrough, 1991 ; Rauses, 1991 ; Steffen, 1990). This molecule has similarity with reduced glutathione, a tripetide (g-glu-cys-gly) in structure and having the empirical formula (g-glu-cys)$_n$-gly, where n = 2–11.

Most reports on phytochelatin induction are on Cd induced ones, either *in vivo* or *in vitro* studies. However, lately we have noted its induction in vivo through Pb in common aquatic plant *Pistia strategies*.

The unique molecule has high affinity for metals and through its -SH group, it makes stable metal-thiolate complex. Thus free metal ions get bound and sequestered. Detailed insight into the mechanisms of this molecule to inactivate metal ions are yet to be determined, they do provide a means of chelating the metal on its entry.

Since induction of phytochelatin in highly specific for metal stress, use of this as 'biomarker' for heavy metal pollution has been highly solicited. This metal binding capacity is not only species specific of a plant genus, but also genotype specific, has been shown in different populations of the perenial grass *Agrostis capillaris*.

Cell Immobilization as a Tool in Waste Treatment

Biotechnological processes, involve the use of biocatalyst either in form of specific enzymes or whole cell organisms. The same holds good not only in industrial processes, but also in arena of environmental management. Immobilization of such biocatalyst, the microbial cell system and possibility of their repeated use has opened new avenues in pollution abatement. Either the naturally occuring biodegrading microbes or their genetically improved counter parts can be maintained in substantially 'unchanged' form to have continued expressions of biological activity and effectivity inside bioreactors through this novel technique.

Advantages of Cell Immobilization

The immobilization cells offer advantages over free cell suspensions, which lie in (1) recovery and reuse (2) avoiding wash out of cells during operations, (3) protection against extreme pH and temperature of medium and toxic pollutants (4) possibility of easy separation from bathing solutions (5) function of cellular multienzyme systems simultaneously to augment effectivity, where needed against use of single enzyme and (6) to have continuous process operation in biological reactors.

In fact the concept of cell-immobilization is intrinsically linked to continuous operation in upgrading of complex wastes. Several reviews have come up on this.

Techniques of Immobilization

Basically involves fixation of cells or enzymes onto a support system. This can be achieved by entrapment and attachment. Induced immobilization is effected through either physical or chemical means.

Encapsulation may be effected with polyacrylamide or Ca/Na alginate beads, while *entrapment* is done on a matrix of collages or siticagel. The cells remain within a porus polymeric structure. This is most widely used in lab incase of both microbial and higher plant or

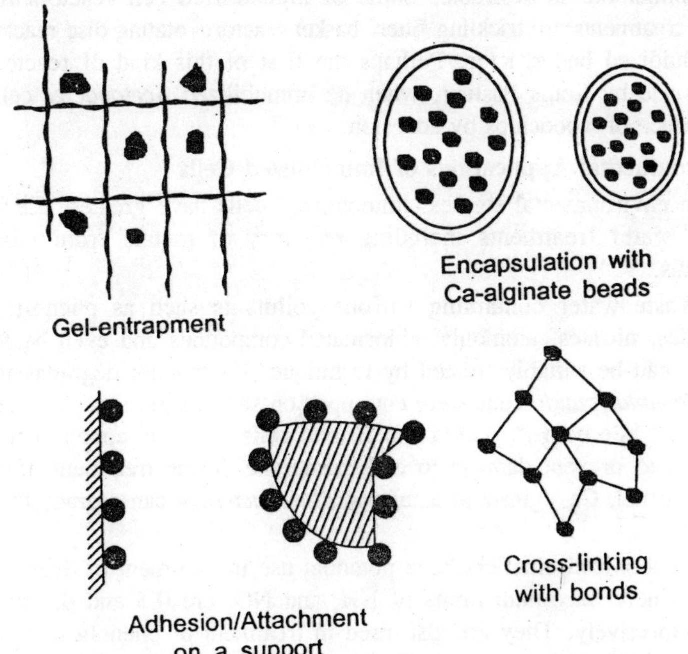

Fig. 12.5. Methods of immobilization.

animal cells from tissue cultures. Alginate is a polysaccharide which when cross-linked by cations such as Ca^+ forms a gel and is highly effective in cell trapping. However, it has low mechanical strength and is sometimes susceptible to biodegradation, whereas polyeurenthanes are a class of non-biodegradable polymers and serve as better cell immobilization agents.

Adsorption of cells on a carrier support is based on concept that microbial cell surfaces are generally negatively charged and need the support which ought to be positively charges. A dense layer of cells is formed and adhere on a solid support. Pretreatment of yeast, *Arthrobacter* cells with aluminium promotes adhersion on glass suface.

Flocculation, on other hand, helps aggregation of cells, as in usually done in activated slude or fluidized reactors, while cell to cell crosslinking is effected with polyelectrolytes or chitosans. This has been done with cells of *Aspergillus*, *Lactobacillus* and animal cells.

For proper development of reactions based on immobilized cells, knowledge of *physiological state* of cells prior to, during and after immobilization, is desirable. Some of immobilized cell reactors for waste treatments are trickling filter, basket reactor, rotating disc reactor and fluidized bed reactor. Perhaps the first of this kind of reactors was made by Louis Pasteur, when he immobilized *Acetobacter* cells on surface of woodchips by adhesion.

Environmental Applications of Immobilized Cells

In environmental studies, immobilized cells have greatest use in waste water treatments including recovery of metals from mine effluents.

Waste water containing various pollutants such as phenolics, cyanides, nitrates, ammonia, chlorinated compounds and even heavy metals can be suitably treated by technique. For phenol degradation, *Pseudomonas putida* cells were entrapped on sintered glass or activated carbon, while both *P. putida* and *E. coli* cells were entrapped in Ca-alginate to prevent damage to cell membrane during treatment of 4-chlorophenol, Ca-alginate immobilized *Flavobacterium* can degrade PCP in soil.

Immobilized nitrifiers have potential use in treatment of drinking water where maximum limits of NH_4 and NO_2 are 0.5 and 0.1 mg/ml. respectively. They are alsö used in treatment of phenolics. They have been used in treatment of excessive fertilizer treated soils and fish hatcheries.

However, several studies have been made in recent years on effective removal of specific pollutants with specific degradative microbes.

Table 12.1. Immobilized microbial removal of some Toxicants.

Immobilized microbes	*Support*	*Substrate*
1. *Cryptococcus elinovii*	Activated carbon	Phenol
2. *Pseudomonas putida*	Sintered glass	Phenol
3. *Enterobacter*, *Alcaligenes* spp.	Ca-alginate	4-chloro-2-phenol
4. *Arthrobacter* spp.	Polyacrylamide	Triethyl lead
5. *Thiosphaera* spp.	Agarose	Ammonia
6. *Rhizopus arrhizus* (fungus)	Polyacrylamide	Cu-effluent

To make the system cost-effective against various contaminants, co-immobilization of microbes of more than one type is to give desirable results. Moreover with development of GEMs and their use of treatment of contaminated waters and soils, stability of plasmids is extremely desirable. Cell immobilization in likely to afford such stability and help in use of such modified microbes.

13

INDUSTRIAL POLLUTION

The contribution of the chemical industry to human welfare is evident in the wide range of chemicals used in drugs, cosmetics, plastics, synthetic fibres, paints, cleansing agents, and many other kinds of consumer products. Unfortunately, most of the chemicals end up in the environment. A high proportion is sluiced down the sewers and ultimately into rivers, lakes, and oceans. Many of the chemicals are highly toxic to living organisms. Some of them find their way into the food chain, disrupting or impairing the natural biotic cycle, threatening populations of organisms with extinction, and jeopardizing the quality of man's food supply. Natural waters from some rivers and lakes are becoming unfit for domestic use and some streams have become biologically barren. The dependence of present day society on chemical technology is almost beyond comprehension.

Synthetic chemicals of every conceivable kind, from medicinet to miniskirts, are part of our everyday life. Nearly two million chemical compounds are known and several thousand new ones are discovered each year. More than 9,000 synthetic organic compounds are used commercially, with 300 to 500 new ones being added annually. When the explorer Thor Heyerdahl sailed his papyrus raft eastward across the Atlantic in 1969, he was dismayed at the abundance of flotsam that he encountered. In some areas the floating debris carried foul-smelling oily material. About twenty years earlier he had sailed across the Pacific in a similar craft the *Kon Tiki*, and had experienced a relatively unpolluted Pacific Ocean.

During the intervening period, relentless pollution with garbage, sewage, industrial wastes and similar debris had established a markedly deteriorating trend in the quality of the oceans. Oceans are not having the rapid turnover that takes place in rivers, springs and lakes.

The oceans have been more stable and maintain their chemical composition over periods of thousands of years. Injurious chemicals can build up in concentration over long periods of time unless they are decomposed. Poisonous metals have been indestructible. A classic example has been the disastrous result of discharging industrial effluent containing, mercury, such as that which occurred in Minamata Bay of Japan. Even some synthetic organic chemicals are having remarkable stability, as, for example, DDT and many of the plastics. Under the anaerobic conditions of silt and mud, chemical substances may be relatively free of oxidation and other chemical degradation as well as resistant to bacterial decomposition. But it cannot be regarded that chemical pollutants will remain indefinitely at the point of deposition.

Some of the pollutants may become buried in the sediment near the coastline where they can be consumed by microorganisms and mud-inhabiting worms. Thus, chemicals and their degradation products are taken up in the food chain and passed from one species of organism to others that prey upon them. Some of the organisms may be carried by ocean currents fish and other organisms may travel under their own power for great distances. A large number of organic and inorganic substances have been identified in domestic sewage.

Materials that are known to be potentially serious pollutants include sundry chemicals from domestic disposal, such as from household cleaners, medicines, and chemicals in food, wastes a wide variety of industrial chemical wastes, including petroleum products, phenols, solvents, chemical intermediates, metallic wastes, and by-products which, may be highly toxic to living organisms; pesticides from agricultural use or industrial waste disposal; and radioactive chemicals, some of which have toxic properties lasting for thousands of years and whose fate in the food chain is largely speculative, though they are among the most toxic substances known. Sometimes contamination of the underground water supply, as well as surface waters, comes from unsuspected sources, as in cold climates where roads are commonly salted during the coldest winter months to lower the freezing point of water and prevent the formation of a hard, slippery coating of ice.

In porous soils, large amounts of salt leach into and through the soil where it can make water unusable. Industry is the largest single user of water, accounting for 50 percent of the daily water requirement. Treatment of industrial waste waters gets complicated by the presence of a wide variety of both inorganic and synthetic organic pollutants, many of which are not readily susceptible to biodegradation. Solvents, oils, plastics, plasticizers, metallic wastes, suspended solids, phenols, and various chemical derivatives of manufacturing processes are apt to be difficult to identify and impossible to remove without more advanced technology than we now possess.

Some of the substances have been known to be highly toxic to living organisms but have been of unknown effect as environmental pollutants. One method used to dispose of industrial toxins has been to inject the waste water deep underground. A large petrochemical plant pumps wastes containing phenolic compounds into a well 6,000 feet deep and wastes containing a class of chemicals called nitriles into a separate well 7,000 feet deep. Disposal of toxic substances deep underground need careful study of the possibility or irretrievably contaminating the underground water Supplies because subsurface water may move for considerable distances underground.

Another damage of underground disposal has been that the pressure might cause slippage of deep layers of rock, resulting in earthquakes. A series of earth tremors in the Denver area were apparently associated with deep underground disposal of wastes. Pulp and paper has been the fifth largest industry and is the third highest in the total industrial use of water. Strenuous efforts are being made to reduce the amount of water used per ton of product.

The rate of reuse has been now about 216 percent, among the highest in industry. Pollution stems from suspended matter and large amounts of dissolved organic substances, the removal of which has been generally difficult and expensive. More than 7,000 gallons of water are needed for each ton of wood pulp produced. Treatment costs come to about 30 to 36 million dollars a year. This would increase to about one billion dollars if the entire industry were to achieve 85 percent B.O.D. (biochemical oxygen demand) reduction and nearly complete removal of suspended solids. The iron and steel industries have been large users of water. By the nature of their operations, they have critical requirements for specialized pollution control facilities. The amount of water needed usually ranges from 20,000 to 50,000 gallons per ton of steel produced, although one mill, by using advanced water conservation methods, has reduced its requirement to less than 2,000 gallons per ton.

Water reuse in the industry is generally about 40 percent of the total throughput. Effluents may run at rates of 10 to 25,000 gallons per minute. The processes of a large steel mill may generate as many as 100 distinct kinds of discharges requiring 5 to 10 separate waste flow systems, two-thirds of which involve chemical, sedimentation, or filter treatment. The principal pollutants have been scale, oils and greases, and miscellaneous chemical wastes. Some industries recycle a large proportion of their water needs, especially water used for cooling. One large petrochemical plant uses 1.5 billion gallons of water per day of which only 10 percent is new water. The industry average is probably 1 gallon of new water in each 3 gallons of water used. There has been increasing interest on the part of industry to reuse its water, not only for reasons of expense and pollution control but also because there is an increasingly limited supply of available water.

Industrial Water Wastes

The importance of industrial water wastes can be realised from Table 13.1, which lists the estimated volumes, BOD, and suspended solids before treatment for industries. The table also includes, for comparison, the corresponding figures for the sewered population of 120 million persons assuming daily per capita production of 0.36 m^3 waste water, 75 g BOD, and 91 g suspended solids. A detailed inventory of industrial wastes would, of course, have to include such important quantities as the amounts of toxic metals, refractory organic compounds, and other water pollutants. In some areas with heavy industrial concentration industrial wastes are many times as great as domestic wastes.

Food and Kindred Products

Wastes produced by the food processing industry-meat and dairy products, beet sugar refining, brewing and distilling, canning, etc.—tend to be troublesome mainly due to their high content of putrescible (decomposable) organic matter, which can result in oxygen depletion and water supply impairment in much the same way as domestic sewage. The ment (cattle, hogs, poultry) processing wastes, come from stockyards, slaughterhouses, packing plants, and rendering plants and contain blood, fats, proteins, feathers and other organic wastes. In many countries, these wastes are being, dumped into the rivers.

The dairy industry produces organic wastes high in protein, fat, and lactose from milk and cheese processing. Whey from the production of cheese is an important BOD source in parts of states with an established cheese industry. The beet sugar refining industry produces wastes of high BOD content, including sugar and protein. Breweries and distilleries produce organic solids containing nitrogen and fermented starches from grain processing and alcohol distilling. The processing of food to produce canned or frozen products leads to enormous amounts, of wet solid wastes. Food processing often involves some danger of infectious disease.

Table 13.1. Estimated Volume of Untreated Waste Water and Mass of BOD and Suspended Solids for Industries and Sewered Population (of 120 Million Persons)

	Waste water (billion m³)	*BOD (10^3kg of O^2)*	*Suspended (10^2 kg)*
Primary metals	16.3	220	2130
Chemical	14.0	4400	860
Paper and allied products	7.2	2670	1360
Petroleum and coal	4.9	230	208
Food and kindred product	2.61	1950	3000
Transportation equipment	0.91	54	—
Rubber and plastic	0.61	18	23
Machinery	0.37	27	23
Textile mill product	0.53	400	
Electrical machinery	0.34	31	9
All other manufacturing	1.70	177	420
Total manufacturing	SO	10,000	8200
Domestic (sewered)	20	3300	4000

Textile Products

Textile mill wastes, generated by cooking the fibres and desizing the fabrics, have high BOD and are quite alkaline requiring neutralization and other treatment. The wastes arise from impurities in the fibre and from the chemicals used in processing. The production of 1000 kg of wool leads typically to 1500 kg of impurities (wool fibres, sand, grease, burrs, etc.) and 300 to 600 kg of process chemicals, with a total of 200 to 250 kg BOD. Cotton processing leads to relatively less total BOD but the waste water may have 200 to 600 mg/BOD.

Paper and Allied Products

Paper and pulp mills, in addition to being notorious air polluters, produce a great amount of water pollution. The effluent, is a complex mixture of the chemicals used in the kraft process stray wood chips, bits or bark, cellulose fibres, and dissolved lignin (woody tissue carbohydrate). About 50 percent of the wood used as input is eventually discarded as waste material. Mill effluent tends to be deep brown in colour and can thus interfere with aquatic photosynthesis it also contains compounds toxic to fish such as methyl mercaptan and paper and wood pulp preservatives such as pentachlorophenol and sodium pentachlorophenate. Sulphite liquor is also toxic to shellfish.

Chemical Industry

A wide variety of water pollutants are produced by chemical plants manufacturing acids, bases, synthetic fabrics, pesticides, detergents, and many other compounds, organic and inorganic. Acid wastes result not only from acid manufacturing plants but also from almost all other chemical plants as well and are customarily neutralized to give a pH of at least 6. Wastes from DDT and rayon manufacturing are especially acidic because a large amount of sulphuric acid is used. The production of the herbicide 2, 4-D leads to dichlorophenol in the water, and the effluent may contain 25 ppm dichlorophenol despite treatment that removes 95 to 98 percent of the compound. The waste waters of the phosphate industry contain elemental phosphorus, fluorine, silica, and large amounts of suspended solids. In recent years mercury waster from chemical plants (as well as other types of factories) have caused great concern.

Petroleum Industry

Oil drilling wastes include drilling muds, saltwater brines pumped out of the well with the crude oil, and some oil as well. Oil refineries and petrochemical plants produce an astounding number of different pollutants. These include hydrocarbons, acids, alkalis, cyanides, numerous sodium salts, phenolic compounds, numerous inorganic and organic sulphate compounds and halogenated and nitrogenated hydrocarbons. Many of these compounds cause detectable tastes and odours at concentrations in the ppb range: ethyl mercaptan at 0.00019 mg/1, isoamyl acetate at 0.9006 mg/1 hydrogen cyanide at 0.001 mg/1 etc. Others cause fish flesh to acquire adverse tastes at concentrations of less than 1 ppm; *o* chlorophenol at 0.015 mg/l, ethyl benzene at 0.25 mg/1, kerosine at 0.1 mg/1, etc.

Coal Industry

Coal mines lead to acid mine drainage, including typically 100 to 6000 ppm sulphuric acid. Coal-preparation wastes from coal washeries contain very large amounts of suspended solids-coal, shale, clay, sandstones, etc.

Rubber and Plastics

Wastes from rubber production are having a high BOD, taste, and odour. Synthetic rubber is prepared from butadiene and styrene in a soap solution and coagulated with an acid-brine solution, and the wastes have some of all the materials used. Odours from rubber plants can give rise to unpalatable water for several hundred kilometers. Plastics manufacture and producing wastes with hydrocarbons and other organic compounds, as well as various reagents.

Metal Industries

Steel mills are producing water wastes from the coaking of coal, the washing of blast furnace flue gases, and the pickling of the steel. These wastes are acidic and are having cyanogen, phenol, ore, coke, limestone, alkali, oils, mill scale, fine suspended solids. Larger mills may recover by-products but smaller mills do not find this economical and may simply neutralize the acidity with lime, which leads to a large volume of sludge. Other metal industries are facing similar problems and their wastes generally have the metal being produced or plated—chromium, lead, nickel, cadmium, zinc, copper, silver, etc. —as wall as acids, alkaline cleaners, grease, and oil.

Other, Industries

Other industries are also producing a wide variety of water pollutants. Leather tannery wastes have high total solids, hardness, salt, sulphides, chromium, alkalinity, lime, and BOD. Polishing of optical glass produces red (from iron) wastes containing detergents and suspended solids that do not settle readily. Laundries have turbid wastes which are alkaline and have organic solids. Radioactive wastes result from nuclear power plants, fuel reprocessing plants, and hospitals and research laboratories using radioisotopes. Soft-drink bottling plants are also producing highly alkaline wastes with high BOD from the washing of bottles, which involves removal of cigarette butts, paper, and other debris left in the bottles by previous users.

14

TREATMENT OF INDUSTRIAL EFFLUENTS

The treatment of domestic sewage has been a comparatively simple procedure because it is virtually wholly organic in nature and very readily biodegradable. The only major difficulty which has arisen in the past has been with some of the many detergents which got disposed of with the domestic drainage water. In most countries there have been now laws which ban the domestic use of non-biodegradable detergents, and in consequence the sewage treatment problem they presented has been gradually disappearing. The question of adequate treatment of industrial wastes has been far more difficult to solve.

Industrial wastes have a far greater variety of impurities than domestic sewage. Many of these have been not only non-biodegradable but also they actively destroy the very bacteria that have been required for the sewage purification processes. If industrial water has to be discharged into public sewers it becomes therefore, absolutely necessary that it be adequately purified beforehand. Often it does not pay a manufacturer to discharge all the waste water from a single process to the sewers; instead he will partially purify the waste fluids from one process, and re-use the water for other processes. Generally such a system makes him to recover certain by-products or other materials for re-use. The problems involved have been found to vary considerably from one industry to another.

Food Manufacturing Industry

The industries concerned have been the rapid-growth industries of fruit and vegetable canning, packing, freezing and drying plants, etc.

It has been estimated that in a country like USA annual production of dried vegetables alone accounts for 20 million tonnes of water—used in vast quantities for washing, cooking and conveying purposes. The food industries have been typically seasonal, and tend to produce their largest quantities of waste liquors in summer, when the levels of rivers have been at their lowest. Dairies which produce milk, cheese and butter, discharge wash water and other wastes. By-products like skim milk, butter milk, whey and dirty milk recovered to be used as animal food-stuffs, but, generally speaking, the purification of effluents with a heavy lactose content requires complicated plant to carry out hot anaerobic fermentation. The costs involved have been for most companies, prohibitively high.

Effluents from some food factories have been sprayed over agricultural land, and this can be advantageous; many of the waste materials have been broken down naturally to give valuable fertilizers. However, this method of disposal of organic waste matter has to be carried out under strict farming control, as the wrong kind of organic effluent can be able to kill of certain plants. Factories which are able to process edible oils and fats generally operate large scale extraction plants. Their effluents, unless, properly treated, are having a high protein, fat and sugar content with a very high biological oxygen demand (BOD) value. In the majority of fermentation industries and in yeast factories 'lagooning' has been successfully used.

In breweries, waste water has been coming mainly from the washing and cooling circuits; only a small proportion of the effluent consists of heavily polluted liquids from presses and filters. Sugar mills, flour mills and starch factories have been facing similar problems. Abattoris discharge water having a high content of suspended organic solids and fats; the high BOD of this effluent has been a function of the blood and other materials suspended in it. Good filtration practice is able to reduce the figure considerably and at the same time produces useful by-products for animal feeding and fertilizer production.

Treatment of Food Trade Effluents

Screening has been used to remove larges solid impurities, and the plant may have to be equipped with a shredder, as for example in an abattoir. Oils and fats could be eliminated by the use of flotation tanks in which these substances could be removed from the top in the form of emulsions: In the case of edible oils, effluents are first acidified. Flocculation is done often after the pH gets altered by addition of acids or bases. This is the practice in dairies where proteins are to be eliminated. As the effluents from food factories have need almost totally organic, further purification of effluents is then done by processes which have been similar to those used for the cleaning of domestic sewage, including aerobic and anaerobic bacterial purification, use of activated sludge, and so on.

Pulp and Paper Industries

Effluents from these processes are having both organic and inorganic matter. The pH has been found to vary widely and the effluents have been often strongly coloured. Flotation processes have been used to recover fibre during manufacture; the effluent is supersaturated with air and passed to an open tank where suspended particles rise to the surface. Filter presses or vacuum and mechanical filters have been used to recover the fibres. The pH has to be modified to encourage other suspended substances to coagulate. Flocculation is done in setting tanks which is followed by sludge extraction. Where the effluent has a high BOD, activated sludge treatment is used. It normally becomes necessary to add {nutritive agents which are rich in bound phosphorus and nitrogen, to permit and adequate bacterial build-up.

Iron and Steel Industry

Due to the large quantities required for steel production, it has become commercially essential to re-cycle most of the water used. Water finds use mainly in cooling, gas scrubbing, removal of impurities and sludge and granulation. For efficient re-cycling water has to be treated as it leaves each separate process. This is mainly done in thickeners designed to reduce overall water losses and to produce a highly concentrated sludge—often having in excess of 500 g of solid matter per litre. Water used for washing blast furnace or converter gases could be purified in scraper type settling and thickening tanks. The tank depth could be as much as 6.5 to 8.5 m, to get maximum sludge concentration.

Mechanical flocculation by sludge recirculation as well as chemical flocculation has been sometimes necessary due to the characteristics of the suspended particles. Additional drying in a vacuum filter or by centrifuge has been generally of advantage. Slag granulation water has been always re-cycled after cooling, clarifying and filtration. Rolling-mill water, laden with dross and scale, could be purified in tangential roughing tanks and scraper bridge rectangular settling tanks, with pressure filters at the final stage. Oil could be removed in the settling tanks by natural flotation.

Pickling workshops discharge acid effluents with a high concentration of ferrous sulphate. The spent pickling liquors could be neutralized with calcium hydroxide or calcium carbonate, using automatic adjustment of pH. Dissolved iron could be removed by injection of air, Which oxidizes the ferrous salts The final sludge is having a mixture of calcium sulphate and ferric hydroxides, which could be filtered off and dried.

Mechanical and Electrical Engineering Industries

The principal source of polluting effluents has been surface treatment processes. Although they have been not large in overall volume they could not be discharged into public sewer because they may be:

(a) Acid, *e.g.*, spent pickling liquors;

(b) Poisonous, *e.g.*, cyanides, chromates, etc, ;

(c) Have reducing properties, *e.g.*, ferrous salts.

Various electroplating and pickling processes need the use of concentrated baths and rinsing tanks. Closed circuit re-cycling of concentrated baths employing ion-exchange agents has virtually eliminated the need to discharge such materials to waste. Re-cycling has been now widely used for bonderizing baths and hydrochloric acid sheet pickling baths.

Methods of Purifying Remaining Liquids

Waters having cyanides have been treated with chlorine or ozone after addition of sodium hydroxide to render them alkaline this effectively oxidizes cyanides. Chromates in solution could be reduced by means of sulphur dioxide in acid conditions. Other acid effluent has been then mixed with the treated cyanide and chromate solutions, and the mixture passed into a large tank where calcium hydroxide has been added. The sludge has been allowed to settle out and could be removed by rotary filters and filter presses. Plant can be made completely automatic so that the liquid which leaves it has a neutral pH of precisely 7 and is having neither oxidizing nor reducing properties. If any of the effluents are having dissolved metals, these could be removed almost completely by using ion-exchange techniques. These have been expensive to operate, but they do reduce the danger of pollution by such poisonous metals as—copper, lead, mercury and zinc virtually to zero. Emulsified oils—particularly from machine shops-have been also a nuisance. Waters containing these have been first of all acidified to break up the emulsion, and then passed into a flotation tank or flocculation and settling.

Chemical Industry

Of all manufacturing industries, the chemical industry—in this context including oil, fertilizer and other concerns—has been contributing most to the pollution of water. The problems have been arising from the very considerable variety of materials produced, and the fact that separation of many of these products from waste liquors needs specialized treatment in almost every case. In addition a very high proportion of the waste liquors and solids produced by chemical works has been frankly poisonous. Often inadequate precautions have been taken in the disposal of very poisonous wastes. It has happened, for example, that cyanide drums have been dumped on public ground. The main methods of treatment of effluent liquors have been the following:

(a) Trace quantities of obnoxious dissolved materials cart generally be removed by ion-exchange processes. This especially applies to dissolved metals like mercury, beryllium and lead, which have been toxic and generally very harmful to all forms of life if released into rivers.

(b) Dyes constitute a major source of contamination of effluents, especially when they get mixed with detergents and organic fluids; similarly, many chemical processes have been producing effluents having phenols, nitrites, aldehydes and alcohols. These effluents have to be treated separately by oxidation or biochemical processes which have been found appropriate for inactivating the impurities in question. They cannot and must not be bulked—often the impurities present in one type of waste liquor interfere with adequate treatment of another. As an example, bacteria needed for the degradation of, say, textile fibres, will be killed rapidly by phenol. Some organic impurities may require to be oxidized, while others will have to be reduced.

(c) If it has been possible to flocculate impurities these could be removed by standard settling and thickening techniques. The sludge could be filtered off and dried.

(d) In the case of acidic or basic solutions which are emanating from such processes as acid manufacture, production of caustic soda, nitrates or explosives, neutralization must be done especially using constant automatic monitoring of the pH value of liquids to get released.

By and large, there have been few problems of effluent purification in the chemical industry which cannot easily be solved by proper application of present-day knowledge; yet the chemical industry has been still responsible for widespread pollution of our environment. The reason has been noted and emphasized, is that existing laws in most countries are either not strict enough or not enforced sufficiently.

Toxic Solid Waste

Toxic solid waste from chemical works often includes the sludge removed from liquids. If these wastes could be indiscriminately dumped on public tips, toxic materials would get leached out and would ultimately reach watercourses. Even the smallest amounts of toxic impurities released in this way can have devastating effects, Recent evidence published on the concentration of mercury in the tissues of marine organisms and fish such as the tuna has been a very good case in point. Many toxic solid wastes could be eliminated by tipping into disused mine shafts.

It has been absolutely essential that before this has been carried out a careful and thorough check has been made that the rock from which every shaft has been constructed has been free from fissures and completely watertight. In general, if a shaft is perfectly dry inside, it has been reasonable to assume that this is the case and unless there is a major earth movement, a subsequent hydraulic connection has been unlikely. If the toxic solid wastes have been organic in nature, incineration has been the best method to dispose of them.

Toxic wastes could also be also discharged at sea. If the point of discharge has been outside territorial waters there is no obligation to obtain any authority from anyone to discharge. Nowadays sea pollution has been a matter of international concern, as many organisms concentrate certain poisonous components, so that eventually the poisonous substances again reach land in the tissues of polluted fish. This is a field in which international legislation has been now long overdue.

Atomic Energy Industry

The nuclear power industry due to its inception and by its very nature, has had to give careful consideration to the decontamination and purification of its effluents. This is, as it must be, one of the main concerns of any 'modern' industry. In the uranium extraction plant, grinding and washing of the ore causes acid effluents, rich in ferric iron, which must be neutralized and clarified. This problem has been encountered in the majority of the mines whatever the nature of the ore.

At nuclear research centres, plutonium preparation plants, and at nuclear power stations, a very wide range of purification techniques has been necessary for the treatment of effluents. On waters working in closed circuit, decontamination has, been done by means of ion-

have been developed for this purpose, one example being Calgon corporation's polydiallyldimethyl ammonium chloride.

The filtration step subsequent to coagulation, can get accomplished by sand, diatomaceous earth, or any of a number of multimedia filter materials that have been developed or through the use of micro screens phosphates can be removed in this way using lime to precipitate them but the cost may run very high.

Carbon Adsorption

Adsorption of tastes and odours is an old process. "Activated carbon" is a porous and highly adsorbent form of carbon with a very large surface, area. In granular or powdered form latter is more efficient (but also harder to handle) it will adsorb many refractory organic compounds dissolved in the water. The carbon must eventually (perhaps once a year) be regenerated 'by heating to about 925°C is an air-stream atmosphere to burn off the adsorbed organic material; under proper regeneration conditions the carbon losses are only a few percent. Pilot plants using columns of granular activated carbon in place of conventional secondary treatment methods have proved successful.

Chemical Oxidation

Waste water treatment can also be accomplished using strong oxidants such as ozone, hydrogen peroxide (H_2O_2) or the free hydroxyl radical (OH). Chlorination with chlorine or chlorine dioxide is also possible. Dissolved inorganic compounds are a problem because they are more common in waste water, even after secondary treatment, than they are in the water supply so that they can easily build up in a water reuse cycle. Several methods exist for demineralization: distillation, freezing, ion exchange, electrodialysis, and reverse osmosis. Distillation and freezing have been apparently not economical but the other methods have been all useful.

Ion-Exchange

Ion-exchange can be accomplished by the use of natural materials (such as zeolite) and synthetic materials (such as ion exchange resins). Cation exchange resins exchange their hydrogen ions for metallic cations in the solution passing through the ion-exchange column, while anion-exchange resins exchange their hydroxyl ions (OH^-) for chloride and other anions in the solution. The resins can be regenerated by treatment with sulphuric acid (for cationic resins) or sodium hydroxide (for anionic resins). Ion exchange is very effective and produces high quality effluents, and it is possible to mix treated water with untreated water to produce effluent of any desired quality. The cost of this treatment method has been fairly high at present. One natural zeolite, clinoptilolite, seems to adsorb both phosphate and ammonium ions and may prove to be of great value.

Electrodialysis

Placing and electrical potential difference across the waste water produces an electric current, making the cations to migrate toward the cathode and the anions to migrate toward the anode. Membranes (really ion-exchange resins in sheet form) permeable to only cations or only anions are used to control the migration of the ions and permit demineralized water to be taken out of the appropriate chambers. Organic molecules are not removed and they can collect on and clog the membranes. Another disadvantage of this method has been that it still leaves concentrated waste water to, be disposed of in some fashion.

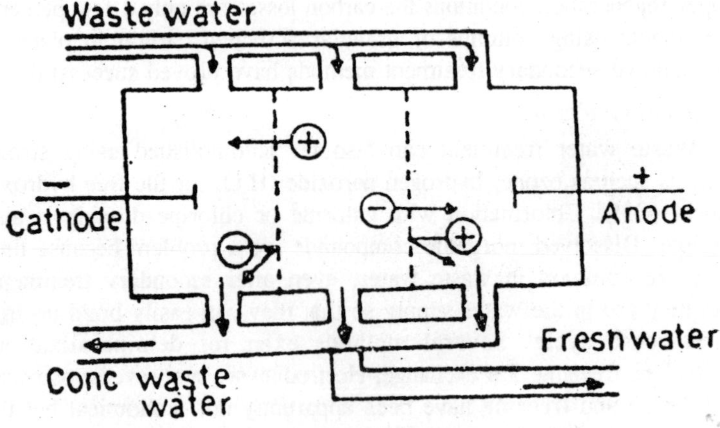

Fig. 15.1. Electrodialysis.

Reverse Osmosis

Another membrane process concentrates the impurities in part of the solution and thereby purifies the other part. Ordinary osmosis involves the movement of water across a semipermeable membrane (i.e., one that is permeable to water but not to the dissolved material) in such a way as to tend to equalize the concentrations Reverse osmosis is the opposite phenomenon produced by applying pressure to the more concentrated solution and thereby forcing water to the other side of the membrane.

Reverse osmosis reduces both the organic and inorganic content of the waste water but some fouling of membranes can still occur. As with electrodialysis there has been highly concentrated waste water to be disposed of. An experimental reverse osmosis plant at Pomona, Calif. led to reductions of 88 percent for total dissolved solids, 84 percent for COD, 98.2 percent for phosphate, 82 percent for ammonia, and 67 percent for nitrate.

Air Stripping

Another process can be used for removing ammonia (NH_2) from waste water. The pH of the water gets raised, usually by addition of lime, and the ammonia driven out of the solution by doing vigorous agitation with air.

Advanced Biological Systems

New biological methods are also being considered for waste water treatment. The use of shallow (perhaps a meter deep) oxidation ponds allows water to be purified by the action of aerobic bacteria and algae. These ponds use solar radiation for photosynthesis and the organic material is used for both bacterial and algal growth, greatly reducing BOD (and also coliform organisms, perhaps through the production of antibiotic substances by the algae).

The ponds may have disagreeable odours if anaerobic conditions are permitted to exist but under proper conditions there is plenty of oxygen produced and the effluent may even be supersaturated with dissolved oxygen. The ponds eventually have to be cleaned out (perhaps every few years) and weeds must be kept under control (this is part of the reason for keeping the ponds shallow).

Costs

The costs of advanced treatment will be some what higher than those of primary and secondary treatment but not very great considering the importance of pure water. The most advanced large-scale tertiary treatment plan is that of the South Tahoe Public Utility District at Lake Tahoe. Calif. It handles 30,000 m^3 (about 7.5 million gal) of waste water daily and incorporates the usual primary and secondary treatment processes plus flocculation and phosphate removal with lime, ammonia removal by air stripping, multimedia filtration aided by synthetic polyelectrolytes, organic removal by adsorption on activated carbon, disinfection by chlorination, and recovery of the lime and activated carbon.

16

Biotechnology of Biodegradation

Biological degradation is considered as a phenomenon of biological transformation of organic compounds by living organisms, especially microbes. The role of microorganisms in the decomposition of sewage and other organic wastes in long known. It has been considered as a natural process in microbial world as carbon and energy source for their growth and takes a pivotal role in recycling of materials in natural and ecosystem. It brings about changes in molecular structure of a compound ultimately yielding simpler and harmless products like CO_2, H_2O, NH_3, CH_4, or PO_3. When compound is not fully broken, it is termed *biotransformation*. Many of recalcitrant substances produced by biotransformation may sometimes be more toxic than the original compound. Such changes are brought about by the catbolic activities of bacteria or fungi by their intracellular or extracellular enzymes, secreted in the medium, biological fate of xenobiotic compounds in the environment can be indicated as shown in Fig 16.1

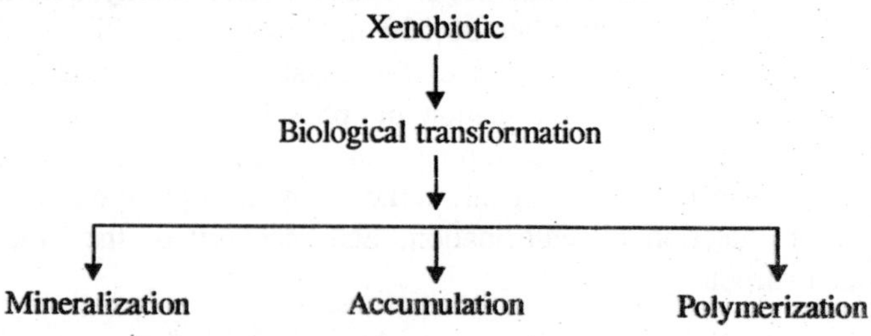

Fig. 16.1. Biological fate of xenobiotic compounds.

Perhaps biosynthetic abilities of living world and the possibility of their degradation by catabolic enzymes evolved parallel but slowly in nature. This has apparently ensured that under suitable conditions all natural organics get decomposed and are not deposited in environment with the exception of natural polymers like lignin and soil humus getting degraded very slowly. This has helped microbes to act as *scavengers* and reduce the pollution load in natural ecosystem. *Bioremediation* of polluted environment capitalizes on activities of aerobic or anaerobic heterotrophic microbes.

The general scheme of such degradation may be represented as shown in Fig 16.2. Here *K* in rate co-efficient and is a function of bio-degradability of organics.

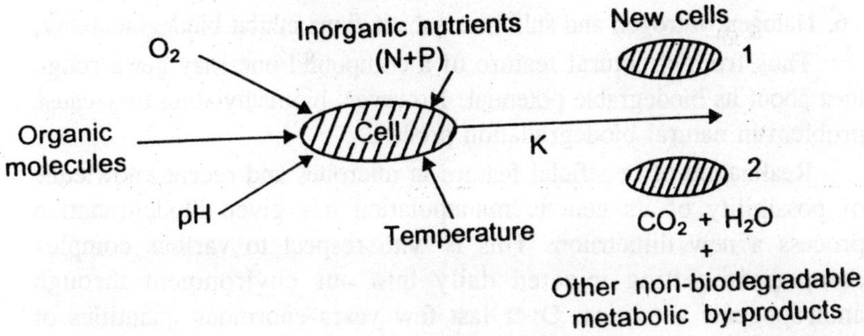

Fig. 16.2. Biodegradation process.

Many of wastes are complex in nature, while strain of microbe may degrade only one type of compound or its related group, for some chemical substances the synergistic action of microbial communities or consortia in a poly culture, displaying wide range of degradative abilities rather than a monoculture, is desirable. Sometimes when the degrading material does not serve as a sole source of carbon and energy for organism, but is associated with another growth substrate, then also gets biotransformed and phenomenon is termed as *Co-metabolism*. For eg., Toluene is primary substrate for *Pseudomonas putida* in breaking *Trichloroethylene*. The insecticide parathion is decomposed by co-metabolism of *Pseudomonas aeruginosa* and *P. Stutzeri*. The controlled environment in a lab condition with one type of organism acting on a single substrate, however, doe not always reflect the real outdoor situation. The factors that affect bio-degradation in situ are temperature, pH, redox potential, availability of nutrients, O_2 supply, biomass of degrades, competition among microbial communities and nature and concentration of the substrate as well.

The chemical nature of compound has great influence in process, which is as follows :

1. Aliphatic compounds are degraded more easily than the aromatic ones. Algae and fungi cannot cleave aromatic rings, whereas bacteria can,
2. Recalcitrance of a compound increase with increased branching, polymerization and presence of polycyclic and heterocyclic residues.
3. Water soluble compounds are easier to degrade than insoluble forms.
4. Alkenes are easier to degrade than alkanes, while alkanes are more amenable than aromatics.
5. For aromatics, degradability may be influenced by molecular orientation, eg., *ortho* > *para* > *meta*.
6. Halogen, nitrogen and sulfonate substitutions inhibit biodegradability.

Thus from structural feature of a compound one may get a rough idea about its biodegrable potential. However, biomethylation may cause problem in natural biodegradation process.

Realization of beneficial feature in microbes and recent knowledge of possibility of its genetic manupulation has given biodegradation process a new dimension. This is with respect to various complex compounds getting injected daily into our environment through anthropogenic activities. Over last few years enormous quantities of synthetic chemicals have been released into the environment, they are posing serious problems being *xenobiotic* and highly dangerous, such as phenols, PCBs, hydrocarbons and other recalcitrant and persistent aromatics. These 'foreign' substances of industrial origin are novel (unknown to nature earlier) to normal microbial enzymatic degradative process. In solving this problem, biotechnology is likely to provide a new and safer approach in early part of this millenium.

EPA list of some organic priority pollutants injected into environment by human activities are :

1. Acenaphthene
2. Benzidine
3. Carbon tetrachloride (CCl_4)
4. Chlorinated phenols
5. Dichlorobenzene
6. Hexachloroethane
7. Naphthalene

8. Polynucleated aromatic hydrocarbons (PAH), namely., Benzopyrine, Toluene.
9. Polychlorinated biphenyls (PCBs)
10. Hexachlorocyclohexane—BHC
11. Pesticides—Aldrin, DDT, Endrin, etc.

The degrading potency of soil microorganisms are being isolated these days from toxic waste sites. The inherent capacity can be enhanced through genetic upgrading of degradative genes. In order to get this, a better understanding of catabolic pathways and genetic bases is essential. Moreover, understanding of the ecological interactions of these new microbial strains is needed for their success in environment.

Till this day about 40-50 microbial strains with suitable degradative potentials for a variety of complex compounds have been isolated worldwide. Some of compounds like organophosphate pesticides, parathion, DDT, 2, 4-D, CCl_4, PCBs, toluene, biphenyls, heavy metals and their organic derivatives been proved to be successful. Agricultural University at Wageneningen has developed a bacterial strain that can break down *venyl chloride*. Recently at University of California at Davis, K. Scow and his group have identified a microbe named PMI, which can degrade MTBE a fuel additive and a potential carcinogen present in soil, in 6 days.

Some of microbes which can degrade various chemicals are:

Chemicals	*Microbes*
1. Hydrocarbons	*Pseudomonas*, *Nocardia*, *Arthobacter*, *Mycobacterium*
2. PCBs	*Pseudomonas*, *Candida*, *Alcaligenes*
3. Phenolics	*Pseudomonas*, *Flabobacterium*, *Trichosporon*, *Bacillus*, *Candida*, *Aspergillus*
4. Polycyclic aromatics	*Arthrobacter*, *Nocardia*, *Alcaligenes*, *Pseudomonas*
5. Napthalene	*Pseudomonas*, *Nocardia*
6. Organophosphates	*Pseudomonas*
7. Benzene	*Mycobacterium*, *Alcaligenes*

Microbial degradation constitute the basic principle of bio-remediation of organic wastes in the environmental clean-up, it involves steps in proper use of indigenous microflora and development of 'enriched' microbes with genetic manipulation for better efficiency. Many biotech companies in the West now market the 'inocula' of such genetically improved microbes. Recent biotech clean-up of Alaskan beaches is the largest application of this emerging technology. The other biotech innovation in this area is development of 'deep shaft' fermentation system by ICI Limited, which is economical with respect to land use and man power in organic waste treatment. It produces much less sludge than the conventional system.

Aerobic vs Anaerobic Degradation

Microbial degradation or transformation of organic compounds may involve either of the 2 processes of aerobic or anaerobic situation, while in some cases it may need both conditions to detoxify some xenobiotic compounds.

Aerobic degradation

In conventional aerobic system, the substrate is used as a source of carbon and energy, and serves as an electron donor resulting in bacterial growth. The extent of degradation is correlated with rate of O_2 consumption, as also previous acclimation of organism in some substrate. Two enzymes primarily involved in process are di- and mono-oxygenases. The latter enzyme can act as on both aromatic and aliphatic compounds, while for the former, only aromatic compounds can act as substrates. Another class of enzymes involved in aerobic condition are peroxidases, which receive attention recently for their ability to degrade lignin.

Anaerobic degradation

This process is wide-spread occurrence and relies on metabolic versatility of mixed microbial populations present in soils or sediments, when O_2 supply is limited. Growth yield of anaerobic bacteria is extremely low due to low energy yields. It has drawn attention these years due to possibility of decomposition of extremely recalcitrant xenobiotics through this process.

Though anaerobic process is slow, needs long retention time and produces H_2S gas, yet it is more advantageous than aerobic one due to its non-dependence of O_2 supply. Thus saves cost of energy of O_2 transfer. Materials like cellulose and fats, which remain unaffected by the aerobic process, breakdown under this situation.

Three temporary ranges are used in anaerobic digestion :

Cold digestion at about 20^0C.

Mesophylic digestion at 20^0–40^0C.

Thermophilic at 40^0–55^0C.

Denitrification, sulphate reduction, dehalogenation and fermentation coupled to methanogenesis may occur concurrently in some soil or

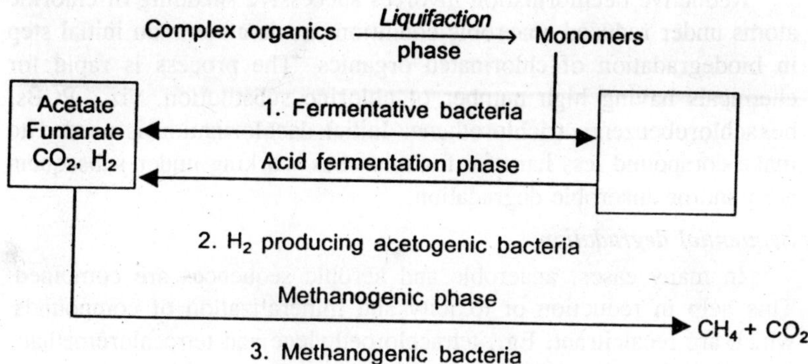

Fig. 16.3. Anaerobic degradation.

sediment as different conditions exist at different micro-habits harbouring consortia of various degrading microbes and their interactions.

Some anaerobic microbial transformation reactions of organic compounds are:

Reaction type	*Examples*
1. Hydro-/dehydrogenations	Phenol, catechol, benzoate, fatty acids, unsaturated hydrocarbons
2. Carboxy-/decarboxylations	Cresol, toluene, benzoate, short hydrocarbons
3. Reductive dehalogenation	Polychlorinated aromatic compounds
4. Dechlorination	PCBs, phenols, chlorinated ethylenes
5. Methylation	Heavy metals

The anaerobic methods of waste water treatment are considered safe, since few toxic chemicals can be stripped into ambient air. Chlorinated xenobiotics need to be dehalogenated to make them harmless and biological treatment in attractive proposition. They need anaerobic situation for dehalogenation by bacterial genera like *Pseudomonas*, *Arthrobacter*, *Mycobacterium*, etc. Unlike aerobic condition, in an anaerobic degradation the chlorinated molecule is used as a direct source of electron. This is exemplified by 3-chlorobenzoate reduction.

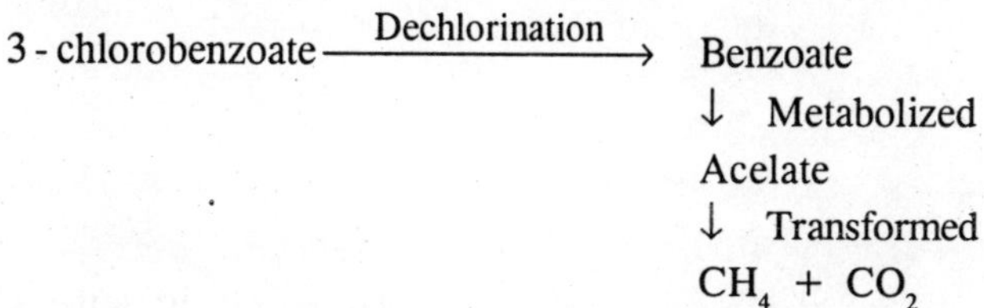

Reductive dechlorination involves successive shedding of chlorine atoms under reduced anaerobic condition and is a common initial step in biodegradation of chlorinated organics. The process is rapid for chemicals having high number of chlorine substitution, viz., PCBs, hexachlorobenzene, trichlorothene. Initial dechlorination is needed to make compound less harmful for microbes working under subsequent aero and/or anaerobic degradation.

Sequential degradation

In many cases, anaerobic and aerobic sequences are combined. This help in reduction of toxicity and mineralization of compounds, which are recalcitrant. Eg., tetrachloroethylene and tetrachloromethane may be mineralized in sequential steps of anaerobic and aerobic conditions, so that initially TCE and chloroform are formed, which, later in aerobic methanogenic stage, are converted into CO_2 and H_2O. In such sequential stages, the BOD reduction is also taken care of as is being done incase of waste water from pulp and paper industries.

For some xenobiotic chemicals, a single bacterium may transform it to another form, but cannot complete the breakdown. In such situation a second group of microbe may act in a complementary fashion to have the total breakdown of compound through a 'team work', this sort of 'synergistic' action prevents the build-up of toxic inter-mediates in environment.

$C_{12}H_{25}$ — *Mycobacterium rhodochrons* → CH_2COOH — *Arthrobacter species* → O — *Arthrobactor* → $CO_2 + H_2O$

Dodecyl-cyclohexane

Bio-oxidation of Phenolic Compounds

Phenolic compounds like phenol, cresol, catechol, resorcinol, pyrogallol and others are released in waste waters from steel industries, oil refineries, petrochemicals, polymeric resins and dye manufacturing units. Structurally some of them are represented below:

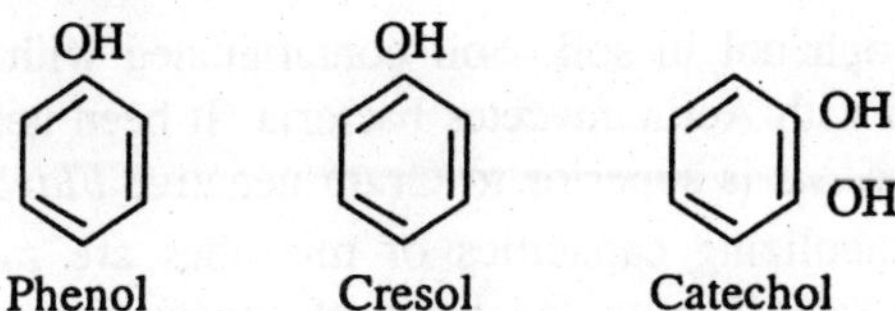

The toxicities of phenolics are well documented in literature. The toxicity may depend on degree of hydroxylation (—OH) and methoxylation of ring. Toxicity concentration of phenol to fishes has been determined at 5.0-33.0 mg/l. They also have strong bactericidal action and thus cause problem in microbial action. Therefore it is absolutely necessary to treat phenol containing waste water before its release and reuse. Normally, recommended levels of phenols in water for domestic use is 1 ppb.

Some of conventional methods of phenol treatments include ion exchange, adsorption on activated carbon, ammonia liquor, soil percolation, etc.

Appreciable bioremoval of phenol from effluent has been tried with microbial 'biofilms' on RBC and through absorption by aquatic plant water hyacinth.

But biotechnologically it is possible to degrade phenolics by biological oxidation with help of phenol oxidizing enzymes or microbial metabolism. It was pointed by number of workers. Organisms endowed with such capacities include a number of bacteria and fungi including yeasts. Immobilization of microbial biomass are in use as a promising biological treatment of ground water phenol pollution. Some of phenolytic microbial genera are:

Bacteria: *Pseudomonas*, *Bacillus*, *Xanthomonas*, *Azotobacter*, *Flavobacterium and Acaligenes*

Fungi: *Aspergillus*, *Pencillium and Neurospora*

Yeasts: *Candida* and *Trichosporon*

Fungi *Pleurotus streatus* and *Trichosporon* or mixed bacterial culture in a bioreactor with residence time of less than 7 hours could degrade phenol. Again Basidiomycetous fungus *Trametes versicolor* can degrade chlorinated phenols. While the bacterium *Pseudomonas putida*, when immobilized on sintered glass were able to degrade 90% of absorbed phenol noted the use of the yeast cells immobilized on polyacrylamide matrix.

Flavobacterium can metabolize 5-chlorophenol as a source of carbon and energy. Alcaligenes when immobilized on granular clay could degrade 4-chlorophenol in soil. Soil contaminated with chlorophenol can be degraded with Actinomycetes bacteria. It been noted that Gram positive *Rhodococcus* is superior to Gram negative *Flavobacterium*.

Phenol metabolizing capacities of microbes are measured from specific growth rates (μ) on the basis of spectrophotometric optical density measurements at 600 nm. It has been noted that catechol is more susceptible to bio-oxidation than Cresols, Hydroquinol and Resoorcinol in same order. Similarly p-aminophenol is more susceptible than 2,4-dinitrophenol, as former is an electron releasing substituted phenol.

Biodegradation of Herbicides and Pesticides

With the advent of *Green Revolution*, there had been a quantum jump in use of synthetic herbicides and pesticides throughout the world to sustain high yielding crop varieties. They have now become a part of modern agriculture. Some of the common herbicides based are—Propham (Carbamate), Dicamba (Aromatic acid), Propanil (Anilides), Simazine, Atrazine (Triazines), Paraquat, Picloram (Pyridines), Glyphosate (Organophosphate), 2,4-D, MCPA, 2,4,5-T (Phenoxyacetic acid). Many of them are highly toxic persistent and show recalcitrance to natural decomposition.

LD_{50} values of some herbicides (in rat)

Herbicide	*LD_{50} values (mg / kg body wt)*
Simazine	5000
Glyphosate	4050
Decamba	1700
MCPA	700

Apart from chlorinated ones, some also have 'surfactants' as adjuvant in their formulations for purpose of retention by leaf surface.

A number of them are now found to contaminate the surface and ground waters through run-off from agricultural fields. Commonly used ones are triazine derivatives, carbamates, organophosphates, aldrin, parquat, diuran, parathion, malathion, etc. Some have been found to be carcinogenic.

Herbicides could be partly decomposed in soils through chemical or photochemical reactions. However, the biodegradability is very much variable. Some of pesticides may appear as recalcitrants, but can be degraded by the process of *co-metabolism*. A number of microbial cell-bound enzymes and extracellular ones can catalyze the breakage of bonds in herbicide molecules. Besides alter the pH of soil and thus help in the detoxification of some of them.

Co-metabolism of MCA & MCPA:

2-MCPA (carbon and energy source) ⟶ *Pseudomonas putida* / *strain PP_3* ⟶ MCA, Glycolate

P. putida as such cannot metabolize MCA. But while metabolizing MCPA, the organisms catalyze dehalogenation of MCA.

Some pesticides like Propanil are partially biodegraded to form Azo-compounds, which may be carcinogenic. Thus biodegradation of such a pesticide may lead to another problem. Again while fatal effect of parathion being known, its molecular modification malathion is less toxic to mammals. Some algae are known to degrade parathion.

Carbamate group of insecticides bring about toxic effects by inhibition of enzyme acetylocholine esterase. Carbaryl is most widely used in this group. It has low toxicity to mammals and gets hydrolyzed to napthol. Degradation of Carbamates are brought about by *Pseudomonas*, *Achroomobacter* and *Flavobacterium*.

Other most used compounds are Decamba (3-dichloro-o-anisic acid), MCPPA (4-chloro-2-methyl phenoxy acetic acid), Diuran (which is a kind of dimethyl urea) and Glyphosate (N-phosphomethyl glycine) and Triazine derivatives like Simazine and Atrazine.

Major advantage of these compounds is possibility of rapid bioodegradation under suitable environment conditions. The chlorinated ones are first dehalogenated and subsequently broken down. The degradation is brought about by *Pseudomonas*, *Azotobacter*, *Bacillus*, *E. coli* and some others. The rate of process would, however, depend on density of microbes and their direct contact, while mobility of compounds in soil layers may reduce the decomposition considerably.

Although DDT is now banned, for its effect on CNS as an organochlorine pesticide, it has however, been reported to be degraded to p.chlorophenyl acetic acid by a number of bacteria, algae and fungi. The filamentous basidomycetous fungus *Phaenerocheate* (lignin degrader) is considered as a versatile one can degrade the fungicide PCP and DDT as well with help of mixture of peroxidase enzymes. It is now used in decontamination of chemically contaminated soil. Biodegradation steps of 2,4-D, Propanil and Diuran have been worked out. For 2,4-D, the steps are as follows:

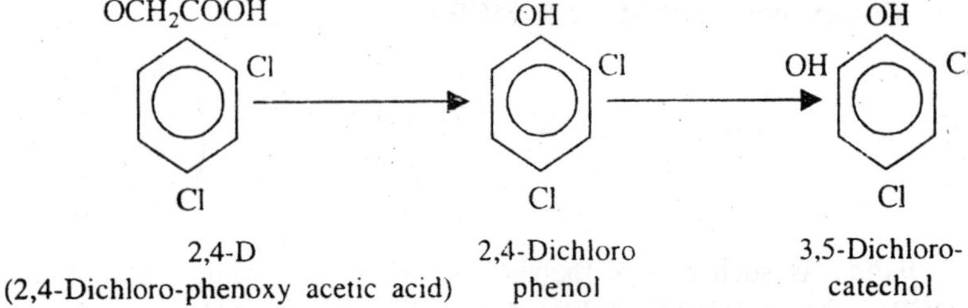

2,4-D (2,4-Dichloro-phenoxy acetic acid) | 2,4-Dichloro phenol | 3,5-Dichloro-catechol

Among the organic mercuric fungicides, the common one is Dithiocarbamate. Some of them may exert toxicity as chelating agents (Thiram), while others (Maneb) decompose to isothiocyanates and creates another problem being a respiratory inhibitor.

Some herbicides may be degraded through the actions of different group of microbes in sequential steps, eg., Dalapon requires atleast six types of organisms to act in complementary fashion. In order to cope with newer herbicides released in environment, microbes also undergo mutation, and in evolution, new enzymes are synthesized to deal with novel substrates. Bioremediation of contaminated soil by *lindane*, a chlorinated pesticide, was achieved by *ITRC*, Lucknow, India, with help of an isolated bacterial strain assigned as EL-1. The microbe could degrade all the major isomers (α, β, γ and δ) of lindane which was not possible earlier.

Biodegradation of Hydrocarbons

The ability of microbes to degrade hydrocarbons was first pointed out by Sable (1950). Interest in this is implicated with abatement of oil pollutions and oil spills, over and above its use in release of oil from oil-bearing substrates (*prospecting*). This is because conventional technologies can yield a maximum of 70% oil from oil-wells. From refinery wastes, hydrocarbon can be degraded by a heterogeneous population of microbes including *Pseudomonas aeruginosa*.

Many bacteria, yeasts and filamentous fungi also shown this capacity. They include species of *Pseudomonas*, *Mycobacterium*, *Corynebacterium*, *Nocardia*, *Arthrobacter* and *Candida*.

Hydrocarbon → Alcohol → Aldehyde
↓
Fatty acid
↓
Further metabolized

Aliphatic Hydrocarbons

Microbial uptake of aliphatic hydrocarbons is impeded by low aqueous solubility. The uptake excretion of emulsifiers for transport across the cell membrane and is regarded as a prerequisite for initiation of biodegradation. It is also largely dependent on availability of molecular oxygen for saturated hydrocarbons.

Anaerobic metabolism of unsaturated hydrocarbons needs hydration of double bond. Both shorter and longer length hydrocarbons as well as branched ones are less effectively degraded.

In general, aerobic process is preferable for remediation of both saturated and unsaturated aliphatics, although unsaturated ones are also amenable to anaerobic processes.

Aliphatic hydrocarbons like n-hexadecanne is reccalcitrant in anaerobic environment, but serves as an excellent substrate for microbes under aerobic condition. The converse is true for chlorinated hydrocarbons. Chlorinated aliphatics include CCI_4 di- and trichloroethylene, methyl chloride and venyl chloride. They cause pollution of aquifers and surface waters. Venyl chloride is used in PVC manufacturing and is dangerous for its potential of inducing leukemia, a fatal cancerous condition.

Aromatic Hydrocarbons

Considerable amount of work has been done on biodegradability of aromatic compounds and the versatility of microbe *Pseudomonas* has been well demonstrated. It can occur under both aerobic and anaerobic situations. Anaerobic breakdown of benzene, toluene, ethylbenzene and xylene (BTEX) has been found to occur, though slowly. This group of compounds has received increasing attention due to their suspected carcinogenicity. Toluene biodegradation steps are shown as follows:

CH_3 → CH_2OH → CHO → COOH → OH, OH

Toluene, Benzoic acid, Catechol

↓

Further degradation

In the degradation of aromatics the first sequence is removal of side chain, while next crucial stage involve opening of benzene ring. It occurs through hydroxylation of the aromatic ring through ortho-or meta-cleavage.

COOH
O_2 Ortho-
OH COOH
COOH
OH COOH
O_2 Meta-
OH

Methanogenic bacteria degrade Benzoate either completely with ring fission or incompletely reduce and decarboxylate. Toluene is reported to be most frequently degraded among BTEX, while o-xylene is reculcitrant isomer.

The presence of p-xylene found to retard benzene and toluene degradation under aerobic condition. Anaerobic degradation of PAH has been described for napthalene and acenophthene in water saturated micro-cosms and under denitrifying conditions. Biphenyl degradation has also been demonstrated under aerobic conditions. However, metabolism of 2 hydroxyphenyl was isomer specific, neither para- or meta-hydroxylated biphenyls could be degraded.

Oil pollution from petroleum hydrocarbons is of common occurrence and recent gulf war impact is classical example. They arise either through seepage from oil fields, blow-out in off-shore oil wells /drilling or from wreckage of tankers. Conventional physico-chemical treatments involve use of adsorbents and surfactants. Microbial biodegradation is known for a considerable time by involving bacterial and fungi, which use these hydrocarbons as food source. It has also been possible to genetically manipulate their degradative pathways to step up detoxifying activity. Often due to deficiency of nitrogen and phosphorus in marine environment, they limit the rate of degradation following oil spills at sea. Oil degrading microbes include *Pseudomonas*, *Bacillus*, *Achromobacter*, *Micrococcus* and *Candida*. Mixed cultures often play a crucial role and co-metabolism is involved in case of complex molecules. Microbes have also been used in oil prospecting. The microbial cells produce surfactants to emulsify oil droplets.

Biodegradation of Some Specific Wastes

Polycyclic Aromatic Hydrocarbons

These include napthalene, phenanthrene, anthracene, and are dangerous compounds originating from oil, tar, wood preserving creosote and fossil fuel combustion. They are degraded slowly. Biotechnological method for remediation of PAH applies *Pseudomonas* spp, and when incubated at 30⁰C, gave optimum degradation result. This helps to develop strategy for bioremediation of PAH-polluted soils and waste streams.

Polychlorinated Biphenyls (PCBs)

They are halogenated aromatics with the formula $C_{12}H_{10-n}Cl_n$, when n = 1 – 10. They have aromatic biphenyl ring substituted with chlorine, viz., Pentachlorobiphenyl.

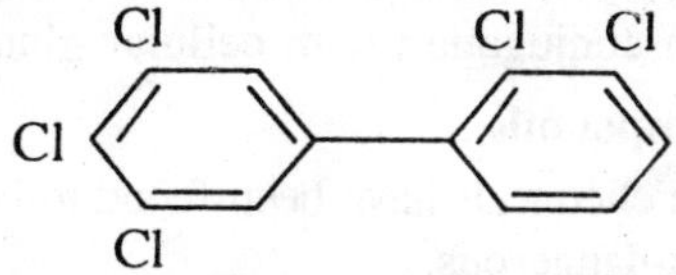

Cl Cl Cl Cl Cl

They are a class of compounds produced either commercially or are found as by-products of combustions, and are used in electrical conductivity, as pesticides, plasticizers, microscopic oils, adhesive, paints, etc. They are industrially value due to thermal and chemical stability, general inertness and resistance to corrosion. Use of PCBs is now dramatically reduced and mostly restricted to transformers. They accumulate in soil sediments due to hydrophobic nature.

PCBs have attracted attention due to their persistance nature, have high bioaccumulation potential, resistance to degradation and have health hazard potentials. They can cause cancer, organ damage and reproductive problems. PCB concentration has been noted to increase with the trophic level, i.e., highest in man and birds. They exposure limit is 1.0 $\mu g/m^3$. They were earlier considered to be undestructible, but now known to be dechlorinated by anaerobic bacteria, followed by aerobic oxidation by members of *Alcoligenes*, *Pseudomonas*, *Acinetobacter* and *Corynebacterium*. To get maximum degrading potential these bacteria must be grown on biphenyl, so their enzymes are turned on. In biodegradation both degree of chlorination and position of substituted chlorine play crucial roles. Compounds with higher number of chlorines are less readily degraded.

The catabolic enzymes in all degraders fall into 2 classes of genes, which complement each other. Genetic manipulation has been found to improve the degrading capacities. The versatile white rot fungus *Phanaerochyte* can also degrade many PCBs.

Synthetic Detergents

Increased use of detergents started in early fifties. They contain *surfactants*, some *builders* and some kind of fluorescent whiteners.

Straight chain compounds are biodegraded, while the branched chain ones are resistant to microbial attack. Biodegradation of surfactants may involve the activity of bacterial plasmides. Biodegradability may be tested in different ways and OECO procedure is widely applied.

Organophosphates (Ops)

They are classed as a group of pesticides, but have some industrial application as oil additives and plasticizers. Unlike organochlorides, they do not generally bioaccumulate, but undergo hydrolysin and biodegradation under suitable environment conditions. Detoxification can occur through conjugation with cellular glutathione (GSH).

Organo-nitro Compounds

Some of these chemicals have been found to be degraded microbially to make them non-dangerous.

(i) *TNT (2,4,6 0 trinitrotoluene)*. It can be tranformed by bacterial and fungal species including *Pseudomonas*, *Clostridium* and *E. coli*.

(ii) *Nitro-cellulose*. This requires pretreatment, i.e., alkai hydrolysis to generate biologically sisceptible substrate followed by anaerobic denitrification.

(iii) *RDX*. It is a heterocyclic nitramine. It can be bio-transformed to formaldehyde and methanol after sequential reduction of nitro group under anaerobic condition.

Vegetable Tannis

They originate as tannery effluents and are toxic to different biota. They are biodegraded by some members of fungi, viz., *Aspergillus*, *Penicillium* and *Fusorium*. Degradation of some tannis have also been reported through the use of edible fungus *Calvatia* (puff ball).

Testing for Biodegradability

It is usually in lab conditions, as it is often difficult to conduct the same in field, though lab conditions may not always simulate the field situations. However, a variety of test methods have been developed, depending on purpose.

Biodegradability of a xenobiotic may depend on or be influenced by a host of factors.

1. Chemical structure of compound and various substitutions.
2. Environmental factors such as the presence or absence of oxygen, pH and temp. of medium.
3. Whether the substrate can provide carbon and energy for metabolism and growth of degrading organism.

So a successful bioremediation of a novel substrates necessitates considerations and attempts from various angles. In many cases, time and efforts may even go waste. So a preliminary idea about the biodegradable potential of chemical concerned may be obtained through some simple techniques and procedures.

1. If structural features suggests its recalcitrance with respect to a particular microbe, it is advisable to check the need of synergistic effects of microbial community.
2. By comparing the disappearance of compound from a biologically active (with 1% $HgCl_2$) growth medium.
3. Decreased UV absorption at 280nm may help monitoring the degradation of aromatic compounds.
4. For specific chemical monitoring of certain compounds like aromatic amines and phenols, diazotization reaction or Folin-Cricalteau reaction may be followed.
5. In certain cases, radiolabelling with C^{14} may indicate the complete mineralization to $^{14}CO_2$ as an indication of biodegradability.
6. Use of indicator reagents in growth media also help detecting (a) conversion of Annilines to Azobenzene by spraying p-anisidine, whereby red-brown discolouration of microbial colonies may occur and (b) for dechlorination-acid-base indicators are used as eosin-methylene blue or bromo-cresol purple. The change of colour indicates the release of chlorine.

Most non-specific analytical methods monitor one of 3 parameters. These are O_2 uptake in an oxidation test (BOD), CO_2 production from complete mineralization of organics and loss of dissolved organic carbon (DOC). The simplest is determination of BOD_5 and its comparison with COD. BOD_5 determines the amount of O_2 required for microbial decomposition in a 5 day test at 20^0C, while COD indicates the amount of O_2 necessary for chemical oxidation. If BOD/COD is more than 0.6, the organic substance is easily biodegradable, if it is between 0.3-0.6, then it points to possibility of biodegradation and when ratio is less than 0.3 it is not bioamenable. Non-specific methods are, however, less sensitive than chemical specific methods mentioned above.

Test guidelines, primarily for screening, have been suggested by OECD and are generally adopted by EPAs in different countries.

Military Application of Biodegradation

Biotech and biodegradation can be used in bioremediation of military toxic waste sites, including decontamination of battlefields.

Many dangerous substances used in various defense activities. They include several cleaning solvents, pesticides, fuels, explosives, heavy metals and propellants. Hydrocarbons are a major group of military wastes and can be biodegraded with suitable microbes like *Pseudomonas*, *Arthrobacter* and *Nocardia*. Normal soil microflora in immobilized form or their enzymes can be used in field bioreactor.

Denitrifying fungus like *Aspergillus fumigatus* has been used as a deterrent to propellants in order to minimize ballistic performance.

The conventional methods of reclamation of battle fields are not only costly and labour intensive, but also generates some other toxic air pollutants through incineration. In contrast to this, bioprocessing offers low-technology, inexpensive, permanent and non-polluting means of doing the same.

17

BIOHYDROMETALLURGY

This is a new discipline of mineral science and is interdisciplinary in nature where microorganism are used in solubilization, prospecting and exploration of mineral deposits. Though concept of 'Biorefining of minerals is known since 1950s, yet during the last 2 decades it has grown into a technology commonly known as '*biohydrometallurgy*'. It is potential application of biotech in mining industry. It has attracted attention for the extraction of metals like copper, zinc, cobalt, lead and uranium from low grade ores. It is not only economic than the conventional processes but also requires less energy and is environment friendly. This technique is now spread in countries like Canada, USA, members of former USSR and India as well for extraction of copper and uranium. The recovery of uranium from low-grades ores through this process is cost-effective for nuclear power stations and in USA, along, approximately 4000 tonnes are extracted per year through this process.

Metal recovery by microbes may involve 2 processes, viz., *bioleaching* or microbial extraction or solubilization of minerals from ores by their metabolic activities and *biosorption* that involves microbial cell-surface adsorption of metals from mine waste waters. The micro-organisms having such efficacy belong to groups of bacteria, algae, yeasts and molds (fungi). Apart from microbes, metal portion potential of higher plants, particularly aquatic weeds, are also being explored for such utilization. Moreover, attempts are being made to use biological colloids in *flocculation* of minerals as an alternative to chemical systems. It has been projected that the worldwide microbial metal recovery may amount to approximately \$ 90 billion by the turn of 21st century.

BIOLEACHING

It is known since long that bacteria are involved in metal ore deposits. They are also known to be associated with corrosion of metals & alloys.

For leaching of ores (heaps & dumps), the leaching solution often contains live microbes some essential nutrients for growth, ferric ion, dilute H_2SO_4, organic acids, proteins, polyaccharides or chelating agents which are metabolic products of bacteria. Bacterial cell-membrane enzyme system also helps in the process.

$$MS + O_2 + H_2O \xrightarrow{\text{Bacteria}} M(SO_4) + H_2SO_4$$

While in indirect process, Ferrous ion is oxidized to ferric and elemental sulphur to sulphuric acid

$$MS + Fe^{++} \rightarrow M + Fe^{+++} + S$$

$$S + H_2O + O_2 \rightarrow H_2SO_4$$

If there is metal oxide in ore, it is solubilized by H_2SO_4 and is called *acid leaching*.

$$MO + H_2SO_4 \xrightarrow{\text{Bacteria}} MSO_4 + H_2O$$

The ore of *dump leaching* is predominantly composed of sulphide bearing minerals, while that of *heap leaching* is oxide bearing.

In the commercial scale, the crushed ore is repeatedly washed with leaching solution. The leached liquor collected from process contains the essential metal, which can easily be separated from sulphuric acid where it has been extracted.

In order to avoid toxicity from products of reaction, the bacteria often remain attached directly to mineral surface. Again in leaching solution a sort of symbiotic activity is noticed, when one kind of bacteria provides one type of food to other and vice versa, thus the food deficiency of one is taken care of by other to maintain the proper growth of culture.

Bioleaching solution contains a number of inorganic salts at an acidic pH with composition of

Components	*Medium*
$(NH_4)_2\ SO_4$	0.05 g
KCI	0.05 g
K_2HPO_4	0.05 g
$MgSO_4$	0.50 g
$Ca(NO_3)_2$	0.01 g

Water	1000 cm^3
H_2SO_4(10 N)	(pH = 3.5)
Fe SO_4	10 cm^3 of a 10%
(Energy Source)	w/v solution

The bacterium most frequently used in bioleaching process is known as *Thiobacillus ferrooxidans*. It is a rod-shaped, motile, non-spore forming and gram negative species. It is merophilic, acidophilic and chemolithotrophic and oxidizes all known metal sulphides to sulphates and elemental sulphur to sulphuric acid. It can also oxidize ferrous ion ($Fe_2{}^+$) to ferric ion (Fe^3 + e) and energy available through this process along with cytochrome enzyme system is used in sulphate production.

Incase of copper ore (*chalcopyrite*), bioleaching is effective and economic. It has been recommended for developing countries as it lowers the cost of conventional mining by supplementing the process activities. Uranium is extracted from uranium oxide (UO_2), where the insoluble tetravalent form is oxidized to hexavalent state which is soluble in acid leach medium.

Leaching techniques have also developed for other metals like nickel, cobalt, silver, gold, arsenic, antimony and molybdenum. However, during bacterial leaching of metals, in many cases, iron present in ore may create problem during further processing. It is essential to remove ion from other metals present in solutions. *Thiobacillus ferrooxidans* can precipitate ion in such cases under aerobic condition.

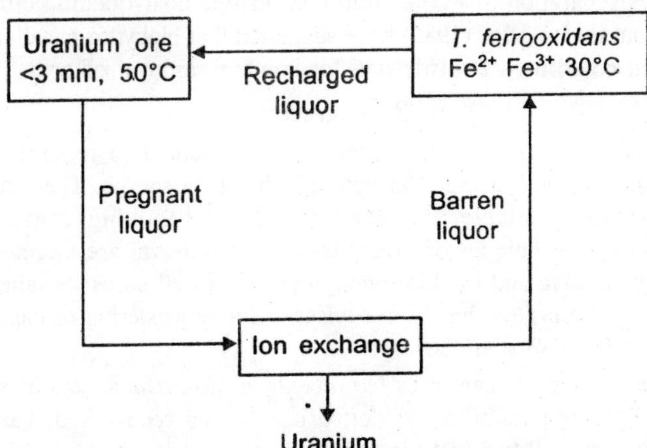

Fig. 17.1. Leaching scheme for uranium ore.

Apart from universally used *Thiobacillus* species, members of other bacterial genera like *Sulfolobus*, *Bacillus*, *Leptospirillum* and *Pseudomonas* have been found to be useful for bioleaching of metals under specific conditions. It has also been found to possible to extract copper and nickel using the fungus *Aspergillus niger* and Gold, by *A. oryzae*.

Bioleaching can also help in removal of impurities from valuable ores and thus upgrade them. Many ores may contain silica impurity. Bacteria like *Rhizobium* and *Bradyhizobium* can remove silica from aluminium ore, bauxite.

It is hopes that bioleaching technology will continue to offer more efficient, cheaper, faster and co-friendly means of extraction of value added minerals. The principal disadvantage of process lies in its relative slowness. Future R & D is likely to solve the problem.

Another possibility of use of bioleaching could be in area of *Bio-desulfurization* of coals with high sulphur content. When this grade of coal is used in thermal power stations, the emission of SO_2 gas can cause high atmospheric pollution problem. Different physical and chemical processes have been suggested and their relative use have been worked out. In recent years, bio-mediated removal of sulphur from pyrite has been considered as a new approach both from economic and environmental points of view. Since its suggestion in 1959 by Zarubina 1959, a number of studies have shown that with the use of *Thiobacillus ferrooxidans* and *T. thioxidans* about 80-90% removal of sulphur could be achieved. Surface treatment with bacterial solutions enhances separation of pyrite from coal in fine coal flotation circuits. In a recent review, Bos (1986) have suggested that biohydro-metallurgical approach could be a realistic one for sulphur removal of coal.

Biosorption

To avoid environmental dangers, it is essential to remove metal contaminants from mine and industrial effluents at source. There metals are non-biodegradable and generally toxic to living organisms. Conventional techniques of precipitation and removal are increasingly getting expensive and not thoroughly efficient in all cases. So alternate strategy of *biosorption* has been envisaged for sequestering or capturing of metals by bio-materials.

Use of micro-organisms as *biosorbents* or *bio-accumulators* of metals offers alternate possibility for detoxification and recovery of valuable but toxic metals. Microbial biomass, either living, dead or immobilized could help the process by way of extracellular precipitation or complexing with cell-surface organic ligands and subsequent accumulation or intra cellular passive accumulation and pumping into special cellular compartments. A large variety of microbes including bacteria, algae, yeasts and molds have shown their metal uptake capacity from dilute solutions biosorbent based granules have been developed for waste water treatment and metal recovery.

Bio-sorbent Microbial Groups

Bacteria

Cell wall composition of bacteria plays a very major role in metal adsorption. Metal deposition on the cell walls have been noted in electron microscopic studies in *E. coli* strain K-12 and *Bacillus subtilis*. *Rhodospirullum* sps. have been found to take up quite bit of Cd, Hg, Pb and Ni. Nakajima and Sakaguchi (1986) have recorded selective accumulation of heavy metals like Co, Ni, Ca, Zn, Pb, Hg and U by various groups of bacteria and actinomycetes. Sahoo 1992 noted that *Bacillus circulans* biomass can adsorb different metals from bathing medium depending on metal concentration and at a pH of 5-6, viz., Cu 80%, Cd 44%, Co 21%, Ni 29% and Zn 10%. Removal of uranium by *Pseudomonas* sps. was reported by Strandberg 1981, Gram negative anaerobic bacterium *Desulfovibrio* was able to precipitate 75% copper from dilute solution. H_2S gas produced during the process could detoxify copper at higher concentration so that bacterial viability was unaffected.

Fungi

The use of fungal biomass, both living and immobilized in metal removal from aqueous solutions has also been demonstrated effectively. The fungal biomass is easily available from fermentation industries. Fungal species that has been extensively used is *Rhizopus arrhizus*. Other genera that have been tried are *Penicillium*, *Aspergillus*, *Neurospora*, *Mucor*, *Yeasts*, *Fusarium*.

For successful application of detoxification of metallic effluents, the fungal biomass, however, needs to be *immobilized* on get matrix to increase the mechanical strength, density and resistance to chemical environment. Immobilization also helps in reuse of fungal biomass after desorption with dilute HCl without loosing binding capacity.

Tobin 1984, noted that *Rhizopus arrhizus* biomass could absorb a variety of different metallic cations and amount of uptake was related to ionic radii. The mechanism of U and Th biosorption by *R. arrhizus* mycelia has been discussed in detail by Tsezor and Volesky (1982, a,b). This fungus, when immobilized on biosorption column of reticulated foam support particles, and under an optimum pH of 7.00, gave better results where negatively charged cells surface ligands helped adsorption upto 2.5 times more than ion exchange. Uranium was also selectively absorbed by *Mucor haemalis*, *Aspergillus niger* and *Pencillium chrysogenum*. *Aspergillus oryzae* could also accumulate cobalt. *Pencillium lapidorum* and *P. spimulosum*, on the other hand, could take up metals like chromium, mercury, zinc, lead, copper, etc.

Tsezos and Deutchmann (1992) checked the validity of an earlier proposed batch kinetic model for uranium biosorption by immobilized biomass of *Rhizopus arrhizus*. It appeared to be somewhat insensitive to effect of Fruendlich biosorption isotherm shape parameter.

Metal uptake by a number of edible mushrooms have also been worked by various workers. Lead and cadmium uptake by *Pleurotus sajor-caju* and mercury uptake by fruit bodies of *Agaricus bisporus* have been documented. While the rate of differential uptake of different metals by the mycelia and sporocarp of edible fungus *Volvariella volvacea* was shown by Purkayastha and Mitra 1992. The order of mycelial mineral uptake was Cu > Cd > Co > Pb > Hg and that of sporocarp was Pb > Co > Hg > Cu > Cd. *Ganoderma lucidum*, a wood rotting fungus, has also been reported to have high metal binding capacity. While the dried and powdered fruit bodies from different bracket fungi are being tried in packed columns for separation of metals from aqueous solutions.

Among fungi, several types of yeasts have also demonstrated their metal accumulation capacity. Strandberg 1981 and Mowll and Gadd 1983 have shown the differential uptake of Hg and Zn, respectively by *Saccharomyces cerevisea*. *Sporobolomyces salmonicolor* and *S. roseus*, while precultured cells of *Crypotococcus albidus* could remove uranium and cobalt.

Algae

They may be either fresh water phytoplanktons or benthic marine community. Information is available on bioaccumulation capacity of metals by members of both groups.

Trollope & Evan 1976 noted concentration of metals in fresh water 'Algal Blooms', different species of single called green algae genus *Chlorella* viz., *C. vulgaris* and *C. regularis* have been noted to accumulate metals like Cu, Hg, Pb, U and Mo. While *C. pyrenoidosa* has been used to recovering metals from sewage sludges. The net like alga *Hydrodictyon reticulatum* has been found to accumulate high amount and Pb. Similarly another green algal genus *Scenedesmus* accumulates Ni and Cu. Mishra 1985 correlated Hg removal from chloralkali wastes to various combinations of blue-green algae. Nature of bondings between metallic ions and algal cell walls was reported by Christ 1981.

Among benthic marine macro algae, members of genera like *Laminaria*, *Fucus*, *Ulva*, *Ascophylum* and *Codium*, have shown great promise in minimizing the metal pollution load of rivers.

Macrophytes

Besides microbial groups, these days some of aquatic weeds have also been found to posses metal absorbing capacity from the mine effluents and help recovery. Well known among these plants are water hyacinth (*Eichornia crasipes*), water lettuce (*Pistia stratiotes*), duck weeds like *Spirodel* and *Lemna*, aquatic ferns like *Salvinia* and few others.

Conclusion

During 30 years *biohydrometallurgy* has made considerable progress in providing opportunities for the extractive metallurgy (*bioleaching*) from low-grade mineral resources. It has also offered the possible of maintaining environmental safety through *biosorption* by microbial biomass. In all those cases, there is a necessity of multidisciplinary approach. There is a need of developing genetically designed organisms for specific functions of high metal resistivity and faster metabolic rates to enhance metal solubilization process. The advancement of present day genetic manipulation techniques may help achieve this end in near future. Indeed, in this respect, a lot of work has been initiated for experimenting with *Thiobacilli*. Other than the microbial sorption of metals higher plants too are being attempted to help recovering metals from mine effluents.

18

BIO-PRODUCTS FOR ENVIRONMENTAL HEALTH

Biotech has been used in sectors of agricultural and energy sources, and helped minimizing the deterioration of environmental health and is possible through development of different novel bio-products.

BIOPESTICIDES

The use of chemical pesticides has increased both in agricultural practices associated with *Green Revolution* and in household demand to control many insects and pests. Environmental implications of these chemicals are well known, some are dangerous and non-degradable in nature. In recent years, there also been development and use of *biopesticides* or *biocides* as they are called. These are pest control agents of biological origin, which are friendly and hold distinct possibility for future use as alternatives to harmful chemical ones. The naturally occuring bio-agents include *bacteria*, *viruses*, *fungi* and *protozoa*, though efficacy of some higher plants have been documented as well.

The most successful bio-control agent is *Bacillus thuringensis*. The principle is a cryastalline protein which is highly toxic to pests of crop plants, and production is controlled by 'Bt' gene. The toxin which is a delta-endotoxin with larvicidal activity causes paralysis of insect larval gut leading to death. It is highly target specific while being safe for other organisms. Moreover another toxin which is a 12 to 120 KDa protein, is got from *B. sphaericus* and is active against mosquito larvae in malarial control. While other species of *Bacillus*, *B. papillae* is pathogenic to several bettle insects. Delta endotoxin gene from *Bacillus* species has been 'cloned' in bacteria like *Escherichia coli* and *Pseudomonas fluorescens*. USEPA (United State Environmental Protection Agency) is reviewing applications to conduct field trials with these genetically engineered micro-organisms.

In recent years the toxin-gene 'Bt' has been isolated and recombinant DNA based products have been produced and approved as safe. Powders have been made from *B. thuringenis* and *B. sphaericus* and are available as standard insecticides. 'Bt' formulations are now used against 90 different insects. Many chemical pesticides manufacturing industries have now realized importance and are focusing on R & D of biopesticides. They believe that in near future biopesticides can occupy nearly 5-10% of the commercial chemical pesticides market.

In future through 'cloning' of herbicide resistance gene from natural weeds to a number of crops and some microbes, it would be possible to avoid the use of many dangerous agrochemicals. Transgenic plants of tobacco and tomato have already been produced which are resistant to herbicide *Glyphosate*. 'Bt' gene has also been introduced into some plants so that they are protected against insect damage, eg., Bollgard Cotton, developed by Monsanto Corporation of America, is now on trials in farmers field in India. Although it comes under the purview of agricultural biotechnology, yet it is considered due to its implication in environmental scenario, as it would help to minimize pesticide use.

Among other groups of bacteria, some species of *Pseudomonas*, have also shown biocidal promise. In use of viral pesticide, '*Baculovirus*' group is noted to be effective in controlling many key pests. The most common among is NPV and its product is now available in USA as *Enclar* and *Biotrol*. In fungal group *Trichoderma*, *Gliocladium*, *Entomophthora* and *Beauveria* are active against many beetles, moths and aphids. The protozoan species that have shown promise for insect pest control being to *Mattesia*, *Lambronella* and a few others.

Apart from micro-organisms, among products obtained from higher plants with insecticidal potential, very few have reached levels with exception of Neem. Neem products are now available in market in many forms. Efforts are being made to improve the neem tree genetically to have better yield of *Azadirechtin alkaloids*.

The oldest plant derived insecticide is *pyrethrin*, comprising 6 close esters. These are natural sources of killing insects. The principal source of this compound is the plant *Chrysanthemum cinerariaefolium*.

Synthetic pyrethroids are long in use. But availability of natural ones is limited. Therefore efforts are on to derive them through tissue culture techniques as 'Secondary Metabolites'. Tissue culture derived pyrethin is more effective than that got from natural extraction. An American biotech company is making efforts to have genetically engineered organism with pyrethin producing gene incorporated into its genetic complement, so that it serve as 'factory' in large-scale culture. On the other, with realization of biocidal property of *Thiopenes*, attempts also being made to get it from *Tagetes* species through tissue culture technology.

With increase of environmental awareness, attention shifted from known agrochemicals like DDT, BHC, Carbamates to biodegradable pesticides and herbicides. These are not only friendly products of biological origin, but less expensive and there is possibility of development of resistant pest varieties. The limitations of bio-pesticides being slow in action and sometimes having unpredictable efficacy, be overcome by future research.

The companies involved in development and marketing of bio-pesticides in US include Ecogen, Mycogen, Monsanto and others. Some of these *bio-agents* are commercially known as Mycostop, Binab-T, Soilguard, etc. Such bio-agents are given the standard 'Eco-Mark' of *earthenpot* indicating safe use. In India, Biotech centre at Anna University, Madras is engaged in indigenous development of biocides. Ongoing biotech and genetic engineering research throughout world is expected to help in future in development of many natural pesticides and weedicides to contain pollution of soil and water. New market opportunities are opening up for bio-pesticides as many synthetic chemicals are banned based on consumer preference.

Bio-Fertilizers

Soil microbes have beneficial effects on crop productions as *bio-fertilizers*, i.e., living organisms used in fertilization of soil. This is due to ability of many of these microbes to 'fix' atmospheric free nitrogen which is abundantly available and useful in supplementing usual application of chemical nitrogen fertilizers and help enriching the soil. The manufacture of chemical fertilizers is not only expensive and dependent on use of fossil fuels, but pollute environment and affects water holding capacity of soil.

Cultivation of legumionous species as a means of enriching the fertility of soil is long known due to ability of fixing atmospheric nitrogen with help of *Rhizobium* present in their root nodules. In nature, apart from different species of *Rhizobium*, a number of other microbes including bacterial species and cyanobacteria are present as nitrogen fixers. Some are free living in soil, while others are associated in a symbiotic relationship in roots or leaves of host plants. Their effective use help avoiding chemical fertilizers and prevent pollution.

More than half a dozen species of *Rhizobium* are associated with root nodules of several legume plants, while *Azospirillum* is with root hairs of grasses and *Frankia* with a number of other shrub and woody plant root systems including *Alnus* and *Casuarina*. Other bacterial genera endowed with such capacity include species of *Klebsiella*, *Azotobacter*, *Clostridium*, *Rhodospirillum* and *Actinomycetes*, which are free living.

Of algal group, 2 known blue-green algae are *Anabaena* and *Nostoc*, which not only occur in free living condition in soil, but *A. azollae* also come in symbiotic association in leaf-packets of free-floating aquatic fern *Azolla*. Now *Azolla* is commonly used as an organic input in rice fields. It is low cost biofertilizer and can add upto 40 – 60 kg N per hectare per crop.

The biological mechanism of conversion of atmospheric molecular nitrogen to ammonia is influenced by key enzyme nitrogenase and its work is genetically controlled. The gene which confers specificity of N_2 fixation has been designed as *Nif* gene, which has actually been found to be a cluster of many gene loci. Genetic analysis of N_2 fixation has been thoroughly worked out in case of bacterium *Klebsiella pneumoniae* which is fairly distributed in soil and water. The *nif* gene of *Klebsiella* has been 'cloned' in other bacterium *E. coli* and in fungus yeast by genetic engineering techniques. Biotechnology in possibility of inducing the capacity of N_2 fixation is non fixing microbes and in higher plants through the protoplast culture and gene transfer technology. It is very complex system of molecular biology and has become a high priority research all over the world challenging biotechnologists. If and when a breakthrough is made, it is going to benefit mankind at large. In India, cone national centre and many regional centres have been established to develop and improve the technology by bio-fertilizer production.

Apart from use of bio-fertilizers, there is a possibility of improving the soil texture with use of organic fertilizers obtained from huge organic wastes and through bioprocessing of huge biomass of aquatic weeds which are normally considered as obnoxious. Late Prof. N.R. Dhar of Shiela Dhar Institue, Allahabad, was a great advocate of this concept. This way we can minimize damage to soil, already done by excessive use of chemical fertilizer, in long run.

Bio-Energy and Fuels

Human civilization largely depend on energy and sources of energy is non-renewable fossil fuels coat and mineral soils, besides forest woods. However, continued use of resources are not only depleting their finite sources, but putting ecosystem under great pressure of pollution hazards. This led to turning to renewable alternate sources of energy of biological origin. They would not only supplement conventional sources, but also provide a cost effective means of generating cleaner environment. Biotechnological advances in area of energy supply through channelization of bioprocesses in becoming an economic and environment friendly reality in future.

Biotechnological strategies for 'organic synthetic' fuel generation are based on microbial processing of plant bio-mass to yield maximum fuels like methane gas, alcohols, as also generation of hydrogen gas through the manipulation of physiology and metabolism of some groups of plants and microbes as well.

Biomass Resources for Fuel Generation

Among possible bio-fuel sources, as per 'photosynthetic model' comes large scale generation of biomass, where solar energy gets trapped as chemical energy in organic matter synthesis.

$$CO_2 + H_2O \xrightarrow[\text{Solar Energy}]{\text{Chlorophyll}} (CH_2O) + O_2$$

This energy source, especially in tropical and sub-tropical countries, holds high promise of suitable exploitation. In this age of energy crisis and atmospheric pollution, biomass offers a good renewable energy

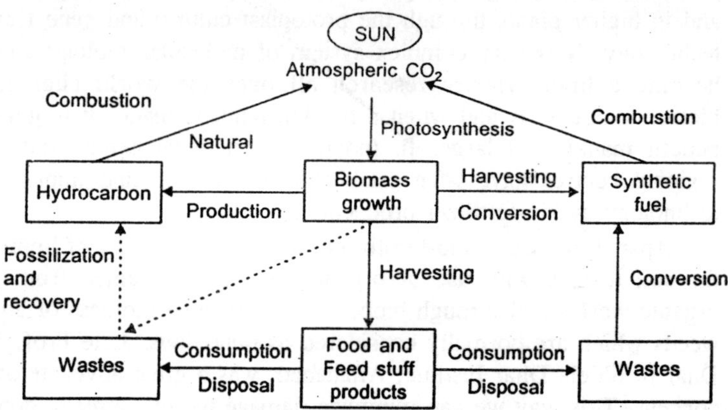

Fig. 18.1. Biomass-to-energy technology.

source. Due to process of recycling no new CO_2 gets released in atmosphere, which thus help reducing 'green house effect'.

There are three main sources of biomass viz., (a) harvesting natural forests, wood lands and aquatic weeds, (b) proper exploitation of huge agricultural residue, sewage and organic wastes c) growing energy rich crops. They represent practical source of carbon of non-fossil origin.

Biomass Strategy

The use of biomass as a direct energy source has long been appreciated in many tropical but less industrilized countries including India. By taking recourse to biotechnological methodologies towards large scale propagation of fast growing fuel wood trees like *Eucalyptus*, *Acacia*, *Populus*, *Leucena*. *Albizia* and other through tissue culture and because of possibility of their genetic upgrading, the biomass scenario is looking up. It, however, needs proper conversion technology for turning biomass into gaseous or liquid form. Apart from thermo-chemical gasification, it is possible to gasify the biomass through anaerobic microbial digestion. This excellent source of energy has still remained untapped on a large scale.

The source of fuel could be in form of biogas a mixture of CH_4 and CO_2 or alcohols. These are obtained through suitable microbial fermentation of carbohydrate contents of biomass comprising cellulose, hemicellulose and lignocellulose.

Lignocellulose component of woody biomass is rather difficult to breakdown and nees expensive pretreatment before microbial degradation. Otherwise, lignin can inhibit microbial growth. It has been found in

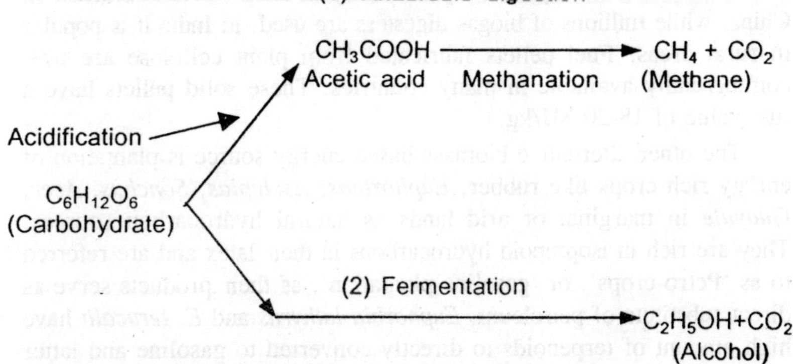

Fig. 18.2. Conversion of biomass into liquid and gaseous fuels.

recent years the white rot fungus *Pleurotus* can break lignin due to presence of 'ligninase'.

Typical fuel values of selected biomass are given in following in the following columns:

Crude Oil	48 MJ/kg
Seed Oil	39-40 MJ/kg
Wood	19-21 MJ/kg
Fibber	19-21 MJ/kg
Algae (*Chlorella*)	27 MJ/kg

Biogas as Energy Source

Biogas is essentially composed of CH_4 (60%), while other like CO_2, H_2S and NH_2 constitute the rest. These are produced during microbial anaerobic digestion of various organic wastes. Two types of bacteria work on process in a biogas plant. The first group in acid forming, eg., *Desulphovibrio* and others. While second is methane forming bacteria, who breakdown acids into CH_4 and CO_2 size, *Methanococcus*, *Menthanothrix*. The slurry, after digestion on drying can provide a good organic fertilizer helping in soil stability.

Water hyacinth (*Eichornia crasipes*) a notorious aquatic weed with a huge biomass, in considered as a good source of methane generation. Its high C/N ratio, low lignin and high moisture content favour CH_4 production. Similarly, another aquatic weed of tropics *Azolla*, is normally highly prized as an agricultural fertilizer due to its symbiotic N_2 fix by alga *Anabaena azollae*, also yield equivalent CH_4 as from water hyacinth.

Production of biogas as fuel has received boost in different parts of world during energy crisis, particularly in third world countries. In China, while millions of biogas digesters are used, in India it is popular in rural areas. Fuel pellets fabricated from plant cellulose are now commercially available in many countries. These solid pellets have a fuel value of 18-20 MJ/kg.

The other alternative biomass based energy source is plantation of energy rich crops like rubber, *Euphorbias*, *Asclepias*, *Sonchus*, *Acer*, *Guayule* in marginal or arid lands as natural hydrocarbon sources. They are rich in isoprenoid hydrocarbons in their latex and are referred to as 'Petro-crops', or 'gasoline plantation', as their products serve as direct substitute of petroleum. *Euphorbia lathyrus* and *E. terucalli* have high amount of terpenoids to directly converted to gasoline and latter species can yield an estimated 5-10 barrels of oil/acre/yr.

Alcohol as Fuel

Production of alcohol from starch and sugar by microbial fermentation is long known.

$$C_6H_{12}O_6 \xrightarrow{\text{Yeast}} \underset{\text{Ethanol}}{2C_2H_5OH} + 2CO_2$$

Industrial alcohol is now obtained largely from petrochemical processes. However, its production from microbial sources still remains a viable proposition, particularly in developing countries where large amount of organic matter could be beneficially used for purpose. Crops like sugarcane, sugarbeet and cassava are used to derive ethanol and also methanol to a lower extent.

Since realization of fuel value of alcohol, it has been taken as a good fuel. It is used singly or mixed with petroleum. It has been success in Brazil, where pollution profiles in cities have been minimized due to non-polluting nature of such automobile exhaust. This *green petrol* technology adopted in Brazil now attracted attention of different countries as well. In Germany, automobile engines are partially modified to run cars with alcohol derived from potatoes. In America and Sweden, lignocellulosic material is suitably stripped off the lignin content, so that free cellulose is converted into alcohol for use as automobile fuel substitute.

Sugarcane bagasse, which is normally considered as a polluting waste and poses disposal problem, can also be good source of alcohol production by treatment with themohilic-alkalophilic bacterium isolated from white-ant mound by microbiology department of JNU, New Delhi.

$$\text{Sugarcane bagasse} \xrightarrow[\textit{action}]{\textit{Bacterial}} \begin{cases} \text{Alcohol} \\ \text{Cellulose fibre} \longrightarrow \text{Gas} \end{cases}$$

However, using only alcohol in automobiles, an 'additivate' called *Alcolita* is needed, which can though increase the N2 content in the exhaust. This needs to be tackled properly. It may be mentioned for conversion of sugar to alcohol, immobilization of yeast cells have been found to yield more. Apart from common yeast *Saccharomyces cerevisae* alcohol yield from *Zygomonas mobilis* is better (95%) (Bringer and Sham) while *Candida* spp. can ferment molases.

Biological Hydrogen Generation

Consideration given to hydrogen gas as a clean and renewable energy source. Biological production of hydrogen is feasible by way of photolysis of water by photosynthetic bacteria and different microalgae or by manipulation of nitrogen metabolism of legume crops.

Many algae and cyanobacteria, including *Chlorella*, *Scenedesmus*, *Microcystis*, *Oscillatoria* and *Anabaena* can also metabolize molecular hydrogen.

$$H_2O \xrightarrow{\text{Photolysis}} O_2 + e^- + H^+$$

H^+ can be converted to H_2 Gas.

$$H^+ + H^+ \xrightarrow[\text{enzyme}]{\text{Hydrogenase}} 1 \text{ molecule of } H_2 \text{ gas}$$

The hydrogenase enzyme helps in recycling the H_2 gas and prevents it from being released in free form. Thus, if action of hydrogenase can be inhibited by creating O_2 pressure, then free H_2 gas will be released. Some bacterial hydrogenase enzyme also used to catalyze H_2 production *in vitro* from water and light using isolated plant chloroplasts.

$$\text{Isolated chloroplasts} \xrightarrow{\text{microbial hydrogenase}} H^+ + e^- \rightarrow H_2 \text{ gas}$$

(Solar energy induction)

Photosynthetic bacterium *Rhodospirillium* had also been used to generate H_2 gas from organic wastes. Microbialy produced H_2 gas can be used in ***biological fuel cells***, for electricity generation, though the efficiency is very low.

In legume crops, the nitrogenase enzyme reduces N_2 to NH_3 and H_2 gas is released. Normally this gas is lost in soil as a by-product of nitrogen metabolism. It has been estimated that from soyabean crop alone this H_2 generation from a hectare of field could be upto 30 b m^3 annually.

The unicellular alga *Botryococcus* could be a good source of hydrocarbon and its culture can provide fuel directly. While purple membrane of *Halobacterium* can serve in photovoltaic bio-cell for generation of bio-electricity.

At tertiary level of petroleum recovery from oil wells, surfactants and polymers are used to get the residual oil. In such cases microbial polymers such as *xanthan gum* obtained from large scale fermentation of specific bacteria can be of much help in release of Trapped oil.

These concepts are at nascent stage of development, future biotech scale up can augment same and turn it into an economic reality.

BIODEGRADABLE PLASTICS

Plastics are part of modern day living, from pack to toys and many other useful items. They are petroleum products where alkene oxides are polymerized to form plastics such as *polythene*. They are non-biodegradable being novel to environment and microbial metabolism is not capable to decompose them. They are dangerous, especially the thin coloured ones. They release toxic chemicals which cantaminate food items, some may even be carcinogenic, more so recycled ones. That is why in western countries manufacture of plastics is now permitted only from biodegradable materials. In India too legislation is under way to ban use of plastics for packing food items. Only virgin transparent materials of certain thickness could be permitted.

The enormous amount of discarded plastics are now constitute a bulk of urban solid wastes endangering the ecosystem. Recovery of such plastics is a formidable task. There is a possibility of recycling the same. There are energy advantages in recycling plastics. The packing plastics are mostly thermoplastic in nature and can be reheated and reformed into some other structures. By *Pyrolysis*, they can be converted into some liquid forms, yielding HCl. But, recycling work itself releases toxic gases in air.

If some biopolymers used in plastic industry, are likely to serve as growth substrates for microbes. It would be eco-friendly with possibility of degradation by soil microbes within a specific period. In this context, it can be mentioned that if sugar is used as a raw material in industry, then either enzymatically or through microbial (*Methylococcus capsulatus*) action, alkene oxides can be produced for further polymerization to plastic. The bacterium *Alcaligenes eutrophus* grows on various carbohydrates and contains polyhydrobutyric acid upto 80% by weight. This can be processed to produce biodegradable plastic. It has been marketed by ICI (UK) as BIOPOL. But production cost is very high. The Tuber Crop Research Institute in Kerala, India,, is said to have developed a starch based technology for bio-degradable plastic. It is on its way to undergo commercial production and may soon hit the market. It has been termed 'green' plastic. Cargill & Dow chemicals in USA has converted plant sugars into PLA and is expected to be commercially produced soon.

A different approach is being made for *bioplastic* production in west. Scientists are trying to produce 100% biodegradable plastics like PHAs and PHBs of vegetable origin. It has been ascertained that production of these compounds are genetically controlled by four genes. Two of these genes are present in many plants, while the other two of bacterial origin, viz., species like *Alcaligenes eutrophs*, *Bacillus magaterium* or *Pseudomonas olevorans*. It is possible at laboratory level to transfer two bacterial genes by gene-technological methods to plants like potato and *Arabidiopsis* at Carnegie Institute at Washington and Michigan State University. This could be exploited in future for large scale industrial production of plastics by simply harvesting these transgenic plants. If it is achieved, in future, it would be much cheaper and readily biodegradable due to vegetable origin. Plant scientists in two corporate sectors like Monsanto & Metabolix, are separately working on a plastic grown in plants that would be ready for farmers as early as 2002.

19

Environmental Management

The science of ecology seeks to explore the relationship between living things and microbes and their surrounding environment. It is the interaction between living (*biotic*) and non-living (*abiotic*) components of environment, which controls the pattern of distribution and abundance of a particular group or community.

Due to over-exploitation and manhandling of nature, present day world is experiencing an ecological strain. In the last few decades ecological imbalance has assumed an alarming proportion, which calls for early restoration through proper management practices. This problem is not restricted to a particular region or country, but also has become a global issue. This is because geographical barriers do not bind water and air. As a result, this leads to the development of the concept of *ecosystem*.

An *ecosystem* can be defined as a broad area comprising the whole *biota* or biological community in relation to the physical or *abiotic* environment of the same, namely forest, grassland or savanas ecosystems. Even though the term was coined by Tansley as early as

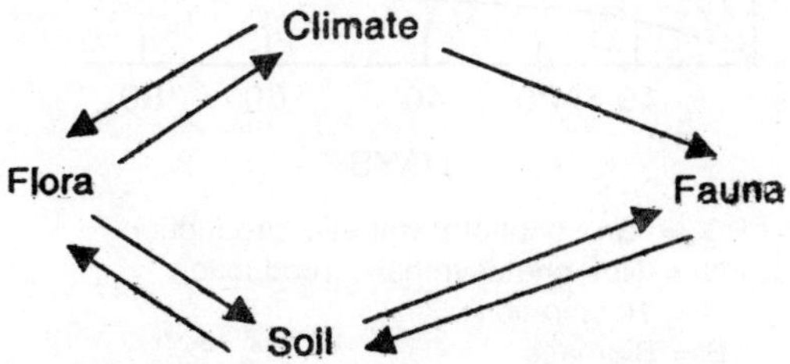

Fig. 19.1. Constituents of an ecosystem.

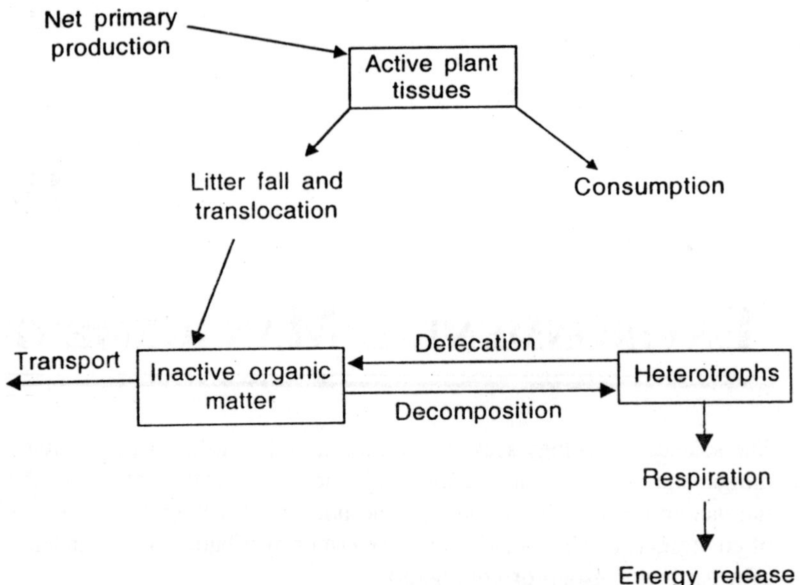

Fig. 19.2. Three-compartment model of an ecosystem.

1935, it was not clearly understood till the middle of the last century. The different ecosystems of the whole world, extending from the earth's crust into the surrounding atmosphere which contains living organisms, constitute *biosphere*. An ecosystem has a definite structure and function.

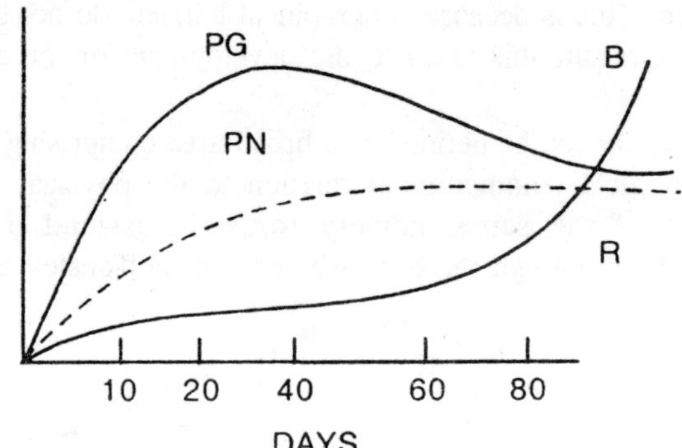

Fig. 19.3. Forest ecosystem energetics.

The functional aspect includes the release of energy, which is unidirectional through the different trophic levels, called *ecosystem energetics*.

Interference of man can be at different levels of an ecosystem. The *nutrient pool* consisting of chemical elements essential to all life process are in constant exchange between living and non-living components of the environment. These inflow and outflow of gases and minerals follow cyclic ways and are called *biochemical cycles*. Among the diverse natural elements, approximately 30 are normally used in biological systems. Out of them, only five namely C, H, O_2, M and P are normally used in higher proportions to support life and as such are known as *macroelements*. And others like Bo, Mo, Zn, Cu, etc., which though required in less amount are essential to life processes and are referred to as *micro* or *trace elements*. The present of microbes in this recycling of matters is of most importance as their enzymatic activities on the one hand help in the *mineralization* of matter and on the other, they acquire food and energy through these processes.

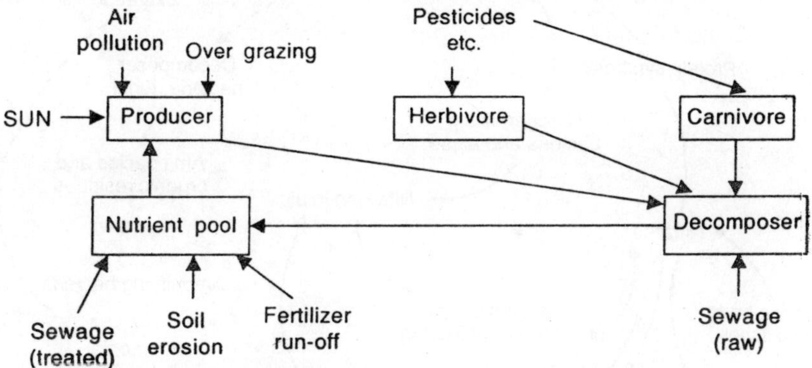

Fig. 19.4. Ecosystem levels affected by human activities.

Man is not only depleting the material resources but also disturbing the recycling process by adverse impact on the microbes. The loss of such elements is difficult for nature to fill up at the same rate. Hydrological cycle are not only influenced by deforestation but also contributes to collection of atmospheric CO_2 resulting in the so called 'green house' effect by raising atmospheric temperature. Usage of ecological phenomena in the support of life processes on the globe is very high and needs immediate conservation of natural resources and their proper recycling. The environmental managers is being drawn to these parameters to have a *sustainable development*, so that a balance is maintained between resource and human utilization.

Biogeochemical Cycles

In the ecosystem, there is a continuous movement of various chemical elements from inorganic to the organic world and back to the inorganic world again. This occurs through the nutrient metabolism by living systems and their subsequent release on death and decomposition. The way taken by an element in such a turnover has been named as *biogeochemical cycle* i.e., the cycles of nitrogen, phosphorous, carbon, sulphur, etc. However, the natural balance in these cycles often get upset through human interference leading to ecological problem. The most suitable examples of this is the over use of phosphorous as a fertilizer in agricultural practices and as important chemical during manufacture of different types of detergents, which enters lakes and ponds through run off and the rapid growth of aquatic flora leading to congestion and slowly turning them into marshes.

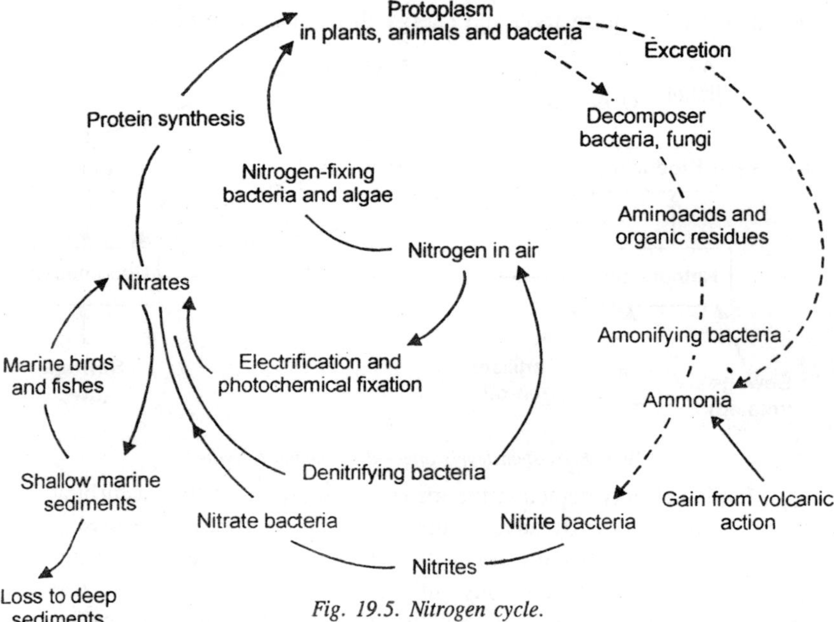

Fig. 19.5. Nitrogen cycle.

Types of Ecosystem and their Developmental History

Ecosystems are grouped as *terrestrial* and *aquatic*. The first one is classified as *grassland*, *forest*, *tundra*, *desert* and others based on availability of water and individual characteristics, while the second is sub-divided as fresh water marine and estuarine on the basis of salt contents of water.

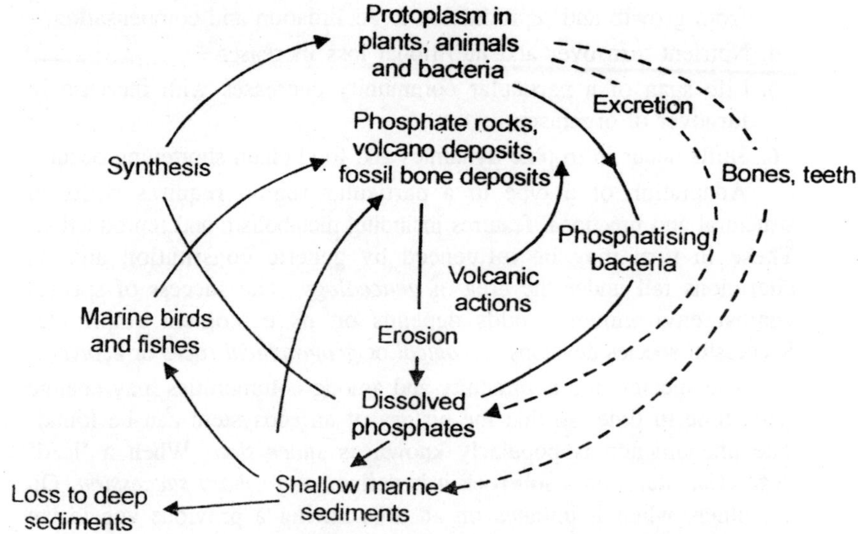

Fig. 19.6. Phosphorous cycle.

Within a large ecosystem, there could be small pockets termed *ecological niches*, which helps in the development of features, which are distinguishable from the rest of the system. High number of ecological niches ensures high diversity. While diversity is a function of time, diversity-stability relationship has also great implications in terms of man's interference with nature.

When a particular ecosystem is controlled by human activities, it is called an *artificial* or *engineered ecosystem*. Organisms that occupy the same type of micro-habitants in different geographical regions are considered to be 'Ecological Equivalents', which help in the extrapolation of data from one region to the other in the formulation of models. The character of whole ecosystem can be altered by environmental stresses both in structure and function. The first is visible by the changes in the mixture of population of different species, while the second is represented by the differences in organic matter, productivity and the rates of gas and mineral release. These are taken as good *biological indicators* of the quality of an ecosystem. A healthy ecosystem is the one which exists through time. According to Cairns et al. (1993), alterations in an aquatic ecosystem may encompass the following parameters.

1. Respiration rate of the community increases.
2. Productivity/respiration ratio gets imbalanced.

3. Ratio of productivity to biomass increases as energy is diverted from growth and reproduction to acclimation and compensation.
4. Nutrient turn over and nutritional loss increases.
5. Life span of a particular community decreases with increase in turnover of organisms
6. Shifts occur in trophic dynamics and food chain shortening occurs.

Adaptation of a type to a particular region requires shifts in structural and functional features including metabolism and reproduction. These in turn may be influenced by genetic constitution and its alterations fall under the area of *genecology*. The success of species against environmental odds depends on its ecological amplitude. Successful species develops *ecological* or *geographical races* or *ecotypes*.

The species in a community and among communities may change from time to time, so that the history of an ecosystem can be found. The phenomenon is popularly known as *succession*. When a 'lead' vegetarian starts on a soil/rock it is defined as *primary succession*. On the other, when it initiates on an area having a previous vegetarian history, which was somehow destroyed by natural upheaval or human activities, is termed as *secondary succession*. Succession on a particular area is named after the physical nature of the area on which it originates namely, *hydrosere* in aquatic area and *xerosere* in dry zone.

An ecosystem of an area is generally found by the dominant or *climax* vegetarian, as it has relatively become stable by successful adaptation to the local variables which are mainly abiotic, examples, the character and composition of tropical rain forests differ completely from those of the temperature ones. Similarly, grasslands of the 2 zones also differ. The different development stages of a particular succession in terms and space called *seral* stages.

Eco-Management

Human progress definitely needs change and development, exploiting nature in this process is not a new factor. In this step of progress, we have over exploited nature and caused harm to its environmental quality. However, the problem has been realized only during the last fifty years when it started to grave and there was a cast and cry throughout the globe. There is a link between development and environmental issues was first raised in Stockholm Congress in 1972 and then at Rio Conference in 1992 as a global forum and has been realized that development is also by maintaining peace in nature. There has been termed as *sustainable development*. It does need a balance whereby the present need may be met without exposing the environmental quality and productivity of nature for future generations.

Then the question arises, what should be the strategy to achieve the goal of simultaneous balance? For this, we need to develop an ethic of survival, a new ecological conscience and thorough understanding of the nature of ecosystems, their sustained resource potential and intelligent management. This can only be achieved by harnessing the services of a group of highly trained environmental managers who would be somewhat different from the commonly trained professional managers and could view the possibility of development without compromising the nature's health. This needs few steps given below :

1. To avoid the polluters and to find proper ways to dispose waste. The slogan should be "polluter pays".
2. Steps to minimize wastes against the backdrop of their environmental impact. This requires specification of 'standards', their execution and constant monitoring.
3. To develop strategies and cost-effective technologies for maintenance of 'ecohealth'. In this respect, apart from the conventional physico-chemical processes, biotechnology can go a long way in tackling the problem.
4. The other strategy which is strongly taken these days is to develop cost-effective technologies for recycling of resources for re-use namely news print, polyurethene, metals from mine wastes, etc.
5. The development of various utility materials of biological origin, eg., biofertilizers, biopesticides and bioplastics, which are all biodegradable by microbial activities and would minimize the pollution load.

In any eco-management study, this health criterion must be taken into consideration before project implementation of an already degraded ecosystem.

In the traditional way of execution of a development project, only engineering aspects based on cost-benefit ratio was considered and not much preference was given on the undesirable impact that the project could have on the surrounding environment. The scenario is now gradually changing. It is now required to have environmental clearance. Environmental impact assessment and environment management studies must be carried out before the sanction and execution of a project plan, be it industrial, mining, dam construction in river valley or urban development.

Environmental Impact Assessment

Assessing the present state of health of the ecosystem where the project would be executed and to work out the possible impact it could bring in course of time. For this, different aspects of air, soil and water analysis along with the ecological features including the composition of flora and fauna is done. It also involves socio-economic studies of the area to gain the first hand knowledge about the possible impact of the project on the local inhabitants. Each feature is given some hypothetical numerical score to have a weighted average of the level of performance, which is called as *environmental quality index* (EQI).

It is necessary to work out *ecotoxicity tests*, which involves biological assessment of the effects of different pollutants. This is mandatory before the release of chemical into the market. At the ecosystem level, the common step to do the same in the *microcosm* tests as model systems having all the pretended environmental features in small scales, so that predictions can be correctly made of the possible pollutant impact. These tests are done on 'standard' biological communities, namely, algae, fish and *daphnia* (insect) for water quality, while earthworm, plan toxicity tests or microbial assessment are considered for soil and air quality.

On the other side, general ecological features of eco-degeneration can be assessed with the recently developed *remote sensing* and *satellite imaging* of the concerned area. It is very sensitive task using airborne sensors and electronic systems and aerial photographic analysis and display of data. Aerial photography also helps in resource and inventory mapping the commonly unapproachable areas including forests and water bodies. It also helps in the analysis of catchment areas, demography and extent of deforestation, if any. This is how it has become a handy tool to have relevant information over a very large area within a short time. Doubtful case area checked later during the field study.

Environmental Management Plan (EMP)

This takes into account continuous monitoring of the health of the ecosystem and safeguard or mitigation measures against the backdrop of adverse project effects if any. It includes mode of waste disposal, waste land reclamation and development strategy.

In undertaking these aspects certain guidelines are being followed, which may differ projectwise, for examples yardsticks would be different for industrial, mining or river valley projects. In all these programs, biotechnology can go a long way in the mitigation of environmental reduction which has been pointed out in the subsequent chapters of this book.

These would be either biosensing the pollutants or their bioassay technologies, and development of more efficient genetically changed biodegrading microbes or pollutant plants. Tissue culture technology also come to the aid of large-scale forestland through micropropagation, i.e., production of number of plantlets artificial/synthetic seeds. This can also be through the development of eco-friendly biodegradable materials of biological origins, instead of generation of xenobiotic toxic chemicals, which are poisoning the world environment, at large.

It has now analyzed that any land use pattern should be based on ecological perspective. No sustainable development in possible without prior assessment of its ecological impact and proper conservation of natural resources including huge biodiversity, which are essential for maintaining proper ecological balance in nature. It has been estimated in 1995 by the well known nature biologist Dr. E.O.Wilson that 'biological species are slipping into non-burning and in the next 3 decades a fifth of the earth's species could vanish for ever jeopardizing natures balance we have to make peace with nature alongside development.

Environmental management is interdisciplinary where urgent co-operation is required amongst environmental engineers, managers, biotechnologists, information technologists and economists to have a fair role in sustainable development. Economics and politics should be put together in environmental governance. To lead a healthy planet, there should be nodal agency like (EPA) to co-ordinate and monitor different activities in environmental arena.

INDEX

E

F

G